# TROUT STREAMS

A MOORLAND RIVER.

[*Frontispiece.*

# TROUT STREAMS

## *Their Management and Improvement*

BY

W. CARTER PLATTS

AUTHOR OF "LIGHT LINES AND TIGHT LINES," "ANGLING DONE HERE," "BETWIXT THE LING AND THE LOWLAND," &C.

*Illustrated with Photographs and Drawings by the Author and Cyril Carter Platts*

THE HARMSWORTH PRESS LTD
8 STRATTON STREET, LONDON, WIX 6 AT

PRINTED IN GREAT BRITAIN BY
HAZELL WATSON AND VINEY LTD, AYLESBURY, BUCKS

To

ELLIS BEANLANDS,

BETWEEN WHOM AND MYSELF, DURING THE LAST THIRTY YEARS, THE CHIEF TOPIC OF CONVERSATION AND CORRESPONDENCE HAS BEEN THE MANAGEMENT AND IMPROVEMENT OF TROUT STREAMS

# Why the new edition

By the Editor of *The Field*

The last two decades have seen two notable developments in trout-fishing, one desirable, the other not. Stillwater fisheries have been immensely increased in number, in extent, and in their capacity to provide sport. If they have not actually been perfected (and time alone will tell whether or no) they have certainly come to constitute a new school of flyfishing with new outlooks and new methods. They have attracted some of the most able fishery managers to exploit them, and some of the finest flyfishers to make the most of them.

Simultaneously, and to the regret of many, the chalk streams have altered the type of sport they provide. Re-stocking, widely employed in order to provide incentives for more rods per mile of bank to rent the rights, and so offset the escalating costs of maintenance, surveillance, taxation and rates, is progressively changing them from natural to put-and-take fisheries. Fewer and farther between are the beats where on high summer days and still evenings an angler can now find the syndromes of season, weather, fly, fish and human skill which, all combined, made up the sport immortalised by Halford, Buxton, Hills, Dunne, Skues and their peers.

But there remains a large body of trout fishermen for whom moving water, natural quarry, and natural fly are all prerequisites to the sport as they wish it to be. Their number would be increased if the available streams had room for them. Hence the decision of the proprietors of *The Field* to include in their re-issue of bygone classics this famous treatise by W. Carter Platts on the management and improvement of trout streams outside the specialised enclaves of the chalk counties. It is based on articles which first appeared in *The Field* as a series.

That there is room for improvement is beyond question. Except in Fen country there are few streams wholly incapable of holding trout, given a little adaptation to that purpose. In traditional trout-fishing country such as Wales, the Welsh March, the Pennines and the West Country, waters already providing sport could provide much more, given the simple engineering works suggested here, all of which presuppose local materials and labour. In Scotland, where the Fisheries Act of 1976 has at last provided a measure of protection for trout, new possibilities have been opened up. Though responsible for overall control of waterways of every kind the Water Authorities, now alert to the recreational aspect of their function, are unlikely to obstruct reasonable improvement for sport.

Despite social and technological change, water in natural watercourses flows as it always did. The possibilities of making them more suitable for trout to live in, more productive of fly, and more attractive to fishermen are the same as ever. The rewards of enterprise are much increased. For landowners, clubs, associations, syndicates and long-term tenants very valuable prospects are opened by the foresight, knowledge and experience gathered here by a pioneer in these matters who has never had a serious rival.

*Wilson Stephens*
February 1977.

8 Stratton Street,
Piccadilly,
LONDON W.1.

## CONTENTS

# ILLUSTRATIONS



And numerous Drawings in text.

# TROUT STREAMS

## CHAPTER I

### MOUNTAIN STREAM AND MOORLAND RIVER

ON nomenclature alone it may be thought, perhaps, that the old formula, " a distinction without a difference," is applicable to the mountain stream and the moorland river ; that the one is so merged in the other as to invite comparison with the case of the medical patient who was so extremely attenuated it was impossible to define with precision where his stomach-ache left off and his back-ache began. As a matter of fact, there is a marked disparity between the two, and the course prescribed for the one, with a view to enhancing its sporting value to the angler, is, in most cases, not the correct treatment for the other.

What, for the lack of a better term, has come to be known in angling literature as a " moorland river " is the main drainage artery flowing down the chief valley of a hilly district with more or less rapidity, and often flanked by alluvial meadows formed on the gravel deposits of countless ages, water-borne or glacier-borne from the neighbouring heights. Of such moorland rivers we have typical examples in the West country, Pennine and Border rivers.

But the term "moorland," as applied to a trout river, has by this time so far extended its scope that it now, to use the epigram of a pedantic friend, "embraces much more than it elucidates." For the name, in the angler's lingo, has passed with the current far from the proximity of the purple uplands, and is now used as conventionally contradistinctive to "chalk-stream," including both upper and lower reaches of trout-bearing lengths of river whereon the use of the dry fly is the exception rather than the rule, and in whose waters it is permitted, in due season and within reasonable limits, to use such lures as the up-stream worm or the clear-water minnow without risk of losing sporting caste. In fact, taking it by and large, the term has in the angling mind taken on the meaning of "wet fly river," just as "chalk stream" has become synonymous with "dry fly water."

These moorland rivers are formed and fed in their upper reaches by the mountain streams—call them hill burns or fellside becks, if you like—that come rioting and rollicking, clattering and chattering, bubbling and babbling with the exuberance and wild abandon of untamed youth down from the ling-clad lofty desolations—down through the scree-cumbered ravines—down through the rocky gorges—down through the dells of tangled hazel scrub—down through the rugged troughs of rough mountain pasture, frequently to gurgle fussily through some grey-walled old-time hamlet just before losing themselves in the moorland river.

The mountain stream, the madcap sprite of the

THE CRADLE OF THE MOUNTAIN STREAM.

[Facing page 2.

wilderness, possesses alluring powers of fascination. Something within us—perhaps a taint of original sin that rebels against hide-bound convention—responds sympathetically to the call of the wild as voiced by this headlong, rackety child of watery nature. And it holds trout! Another fact that adds magic to the call.

I have been told the story of the inquisitive child, speculating upon the Great Beyond, who asked her mother curiously:

"And if I go to Heaven what will I do all day long?"

"Oh, dear, you'll always have a nice little angel to play with," was the reassuring reply. The child pondered silently and solemnly for a little while. Then, yearningly, imploringly, she pleaded:

"But, mummy, dear, if I'm very, *very* good, won't I be allowed to have a little devil to play with now and then?"

As anglers we are—many of us are, at any rate—but children of a larger growth. We may fully appreciate the piscatorial joys of watching the gently poppling float sailing upon the decorous bosom of some carefully ordered and sluice-controlled waterway; or we may exult in the correct cocking of the dry fly on the lazily gliding chalk-stream; but away in that unregenerate corner of our heart we hear at times the echo of the boisterous song of the mountain stream and we long for "a little devil to play with now and then."

I have read many books dealing with the subject of improving our trout fisheries, but I cannot at the moment recall a single one which has treated at any length of mountain streams except from the view-point of the fish-breeder rather than from that of the fish-catcher. Schemes are advocated—and have been put into successful operation—for the formation, at considerable expense, of a series of ponds, where a suitable site can be found for them, beside the stream, and fed from it by a sluice-controlled channel ensuring a steady flow of water through the ponds, while the surplus flood water passes down the original stream. Or you are advised to throw substantial dams across the original bed of the stream to form a chain of pools, connected by a more or less complicated system of drains and sluices, and cut an artificial channel or by-pass in the hillside to carry past the pools all the water not required in ensuring a gentle flow through them. I have in my mind a particular instance where the latter plan was carried out in connection with a commercial fish-farm with great success. But in either case, your little devil has disappeared at this spot, and in its place you have acquired a set of angelic, but very insipid (from the angling point of view), set of docile rearing ponds.

Without detracting from their natural charms the sporting capacity of many mountain streams may be vastly improved, as will be shown in subsequent chapters, at small cost beyond the labour expended on them by a little handful of enthusiasts. Indeed, the alterations effected in many instances add to the attractive appearance of the streams. Before, however, any serious work on a stream is undertaken it should be ascertained beyond doubt that the stream under consideration is suitable for the purpose ; in other words, that the trout will readily take up residence in it. For reasons which are not always apparent, fish will often exhibit a marked predilection for one stream while they leave another severely alone, and it is obviously foolish to prepare quarters for guests who steadily decline to occupy them. An old superstition has it that the trout and the char in Windermere both proceed at spawning time up the combined waters of the two tributaries, Rothay and Brathay, until they reach the point where the streams join, whereupon the trout proceed up the Rothay, while the char unanimously elect to move up the Brathay. Mr. John Watson, who has made a closer study of Lakeland fisheries than any other authority, asserts in his book on "The Lake District Fisheries" that this superstition is inaccurate as far as the trout are concerned, seeing that they spawn indiscriminately in both streams, but he corroborates the statement that the char confine their honeymoon trips so exclusively to the Brathay that the few odd char accidentally taking the wrong turning and finding themselves in the Rothay at once rectify the error by turning back and running up the Brathay. Mr. Watson ascribes the peculiarity largely to some difference in temperature of the two tributaries.

In another instance, concerned with two streams with which I am familiar, the partiality of the trout for one water is very pronounced. The two streams, both typical fellside becks, join forces about half a mile above their junction with the river, and at the back-end a generous crowd of spawners flock up the combined waters. You can watch them for hours on end at one fall, when the water is at the right height and the run of fish at its briskest, leaping at the watery obstructions on the if-at-first-you-don't-succeed principle, often two or three in the air at once. Arrived at the fork, the trout, without any pause to consider the question, pass up the stream to the left with very few exceptions. That stream is a prolific breeding tributary and holds a fair amount of fish the whole year round; the other is almost barren of fish life, and I should consider any attempt to improve it wasted labour until the cause of the trout's dislike for it was discovered and removed. I cannot state the cause. All I know is that the head waters of this unpopular stream ooze out of a big moorland swamp a mile and a half from its confluence with the other stream, which latter emerges from a cave in the limestone rock. Whether we have here another case of the fish being influenced by any difference in the temperature of the two waters, or whether the objectionable stream carries a forbidding sphagnum flavour throughout its whole length, I do not care to hazard an opinion. The moral, of course, is: Don't waste time and labour on a task that promises only failure from the outset.

The work of improving a lowland trout fishery

may, as one disgruntled preserver put it to me, consist of one endless round of damnable reiteration—cut the weeds, net out the coarse fish, stock; then cut the weeds, net out the coarse fish, stock; after which *de novo et de novo* for ever and for ever until the dawning of the Great Millennium. Improving the mountain stream is not at all like that. There is no monotony in it; always some unexpected and interesting problem is presenting itself for solution. Why, on one occasion while we were at work on some dams on the lower length of a mountain stream I was approached, not in the execution of his duty but in a purely private and friendly capacity, by the village policeman, who put it to me, kindly but firmly, that we were acting illegally by depriving the stream of water that ought to flow through the village below, which we had no right to do. He reminded me of the man employed by old Mr. Armistead, of the Solway Fishery. One of his duties was to see that a constant stream of water was flowing through an artificial lake. So it was his invariable practice to proceed to the bottom end of the lake first. Having satisfied himself that the water was running out at that end, he then conscientiously trudged away to the head of the lake to make sure the water was running in at the other end.

## CHAPTER II

### BOULDER DAMS

SOME years ago a friend of mine preserved a delectable upland stream that brawled its way down remote glens deeply cut in a spur of the Pennines. In one portion of its length unaided Nature had so shaped its bed that it presented to me the materialised ideal of my happiest dreams of a mountain trout stream. From dancing pool to dancing pool, down the long terrace, the frolicking water tumbled and cascaded over low natural dams of rock and boulders tightly jammed together in picturesque confusion. The pools being small, some of them no more than five yards in length, some of the larger ones about ten, there was always a good current through them, but also there were slack watery nooks and tiny restful eddies. And the trout loved these rocky terraced pools. Here was every comfort a mountain trout could reasonably ask for—rest for the weary, food for the hungry, animation for the restless. Energetic youngsters played leapfrog in the bustling current; sedate elders retired with their knitting to the shady corners and dozed between meals. Fishing upstream, wading up the middle of a pool and throwing ahead on to the pools above with a couple of flies 7 feet or so apart, I have more than once risen and hooked—and eventually creeled—two trout at one cast in two different pools.

Athletic as the trout of the mountain stream are, they cannot spend their whole existence in battling with a rapid current. They seek resting places. The pools attract them. I have watched sheep-washing operations in summer on the mountain moorland. With boulders, moss and turf the shepherds quickly throw a temporary dam across the stream. It is but a frail structure, swept away by the next freshet, but it serves its purpose of forming a pool of sufficient depth for washing, and for hours that pool is full of hefty moormen wrestling with reluctant sheep, struggling, sousing, splashing, while an ochre-coloured tide pours over the crude dam. Within a few hours of the departure of the men and the dispersal of the complaining flock, you may, if you watch closely, see that pool occupied by appreciative trout.

If the mountain trout want pools—and they do—give them pools. Make home happy for them, and they will show less disposition to desert it for the attractions of the greater river.

There is many a length of mountain stream where the water flows swiftly for quite long distances over coarse gravel shallows studded with half-buried boulders here and there. Feeding trout use the water, but there is little in it in the way of resting place for them. Scarcely anywhere is the water more than ankle-deep, and the only tiny slack spots are the little bubbling eddies behind the upstanding boulders. Having fed, the fish drop back to deeper haunts below, or surge forward in search of harbour above.

Such reaches I always compare in my mind with the far different natural conditions I have described at the commencement of this chapter, and then I begin to see visions of a transformed stream more nearly approaching the ideal. The natural features of the burn force themselves into the scheme. There are sites simply asking for dams to be built upon them. One such is shown in the accompanying photograph. Already nature has laid the foundations for the dam in the shape of a rough line of big boulders, deeply locked in the bed of the stream, while at each end of this line is a great boulder half-embedded in the bank—a valuable asset. More, above, and within easy distance of, the suggested dam site is abundance of material for the structure in the shape of other scattered boulders showing above the water. These may seem, at first sight, of insignificant size, but the boulder in the stream bed possesses, usually, one characteristic in common with the iceberg—there is a good deal more of it out of sight than appears above the surface.

There are other big boulders immediately below the line of the proposed dam. Let them stop there. It is far easier to trundle a three-hundredweight boulder ten or a dozen yards downstream under water than it is to work one of similar size a couple of yards upstream and lift it out of the water on to the dam. Besides, the boulders immediately below the dam, if left undisturbed,

SITE FOR A BOULDER DAM.

[*Facing page* 10.

will probably serve a useful purpose where they are, as will be apparent later.

To my way of thinking—and, to put it with all the impressiveness that colloquialism can impart, " I have been there and had some "—by far the most useful tool for boulder-dam making is the pick. It should be a heavy, strong pick with a stout shaft. The crowbar is a useful implement, especially as an auxiliary to the pick, but it has one serious drawback. The man wielding it, in shifting a heavy boulder in the bed of the stream, spends about half his time in seeking a fulcrum which is not easy to find under water. The pick, in a way of speaking, carries its own fulcrum with it in the rocker-like curve of its metal, which finds a ready purchase on the stream-bed.

To persuade, then, a big boulder out of its long and deep-seated resting-place in the gravel bed, stand just below it and, working over it, drive the point of your pick well down on the upstream side of it until you find that you have got a grip. Then move round to a position just above it, grasp the shaft of the pick and pull upward and upstream, putting all your beef into the effort. Eventually, if neither your pick nor your back breaks in the meantime, the boulder usually exhibits signs of surrender. You raise it a little on the upstream side, and your assistant pops in his bar to hold it in its new position while you get a deeper purchase. In the case of a particularly ugly customer you may have to enlist the services of another man with a pick, or it may even require the united efforts of three men to uproot the enemy; but a bit of plucky doggedness fetches the obstinate rock out of its hole at last. Once out, it is surprising with what comparative ease four hands can trundle a big boulder down the stream, for not only does the current help, but the fact that a considerable part of it is under water lessens the apparent weight. If the boulder is of an awkward shape, with a flat side and wicked angles, it may be necessary to work it slowly downstream the whole of the way to the dam with picks and bars. But there is always consolation in the remembrance that an ugly-shaped boulder will often lock into your dam far more securely than a roughly spherical one, and unless your dam is firmly built it will be demolished by the first flood.

No arbitrary rule can be laid down as to whether you should commence building your dam in the middle or at one end first. In this matter each case should be decided on its merits, with due regard to the best ways and means of jamming the boulders you handle securely against those nature has already fixed in position, and interlocking them with each other. In any case it is best to move all your boulders down to positions near the site of the dam before commencing actual building, and while doing so you may roughly allocate each boulder, by reason of its size and shape, to its particular position in the dam. It is well to have a pretty clear idea of details in your mind beforehand, for it must be remembered that once you start actual construction you are working in rising water during the remainder of the operations.

Having, then, trundled all your boulders near the spot where they will be wanted, you now trundle and coax them a little further, one at a time, until you have them firmly jammed in position and the last considerable gap is roughly closed. Still a big bulk of water is making its way through the chinks between the boulders. You now deal with this by ramming home smaller boulders where you can tightly wedge them, and cramming the crevices with felted moss torn from the stones in the stream, always working on the upstream side of the dam. Then fill up the angle at the upstream base of the dam with small boulders, piling them against the bigger boulders of the dam, and in a few minutes, where a little before was a prattling shallow, you will have an attractive " trouty " pool with a merry cascade over the rough and picturesque dam into what should later be the head of your next pool.

Photography has its limitations. It cannot show you the completed pool without showing the cascade hiding the dam. Therefore, I fall back on pen and ink for showing in one picture the pool formed on the scene depicted in the previous photograph, and the method of construction of the dam.

One point which is shown in the sketch I omitted to mention earlier. As a matter of fact, it had no part in my earliest experiences, and was begot of observing the results of initial experiments. A few stout stakes driven deeply into the bed of the

THE BOULDER DAM COMPLETED.

stream immediately behind the most pregnable parts of the dam add immensely to its powers of resistance to the tearing violence of winter floods.

Here a word of warning may not be out of place. The height of summer, when the streams are at their lowest volume during a period of drought, is the best time for dam-making. Even then the operator is frequently immersed to mid-thigh, and the water of the mountain burn is cold—sometimes very cold—while a blazing sun and the arduous labour put the upper part of the body into a frizzling heat. Under such circumstances I do not, with Scrope, " account it an unpardonable intrusion to place a solution of India-rubber between the body and a refreshing element." On the contrary, such an interpolation is a safeguard against considerable risk. The moorland shepherds of to-day, hard weathered as they are, are wise to the dangers of such exposure, and it is a common thing for them to " India-rubber " themselves in " washing suits " before entering the washing-pool. For dam-making the ordinary fishing stockings will serve admirably, and they are perfectly safe—so long as you do not inadvertently ship a legful of water when bending.

## CHAPTER III

### THE LESSONS OF THE DAMS

In every-day conversation we glibly trot out the hackneyed quotation "When Nature fails then Art steps in," with the easy assurance that with this intervention the happy consummation has been reached and nothing more remains to be done but to celebrate the triumph of Art in a suitable manner with musical honours. Alas, the proverb is but a half-truth. The dam maker on the mountain stream in his novitiate is apt to be slowly, sadly and surely disillusioned, as the truth is borne in upon him that the proverb is incomplete; that, in actual practice, after Art has done its little whack in the improvement line, then Nature butts in and arbitrarily rearranges matters.

Indeed, when Man joins his enterprise with Nature's forces in an undertaking such as I described in my last chapter, the association, in the method of its procedure, has much in common with the literary collaboration of the famous French romancists, Erckmann and Chatrian. The general plan of the work having been decided upon, Erckmann, figuratively, took off his coat and got to grips with the task of materialisation, after which Chatrian read what Erckmann had written and either put it in the fire or made drastic alterations and corrections. So when the enterprising amateur has made his first set of boulder dams, Nature gives them the critical "once over," and proceeds either to demolish or to effect drastic alterations. It is at this stage the would-be improver realises that in selecting the site for his dam it is not sufficient to allow himself to be influenced solely by the fact that a certain spot offers special facilities for the erection of his barrier. He must look ahead. He must, to change the simile, attempt a forecast of how Nature is going to knock the bowling about when she comes in for her second innings, and he must set his field accordingly.

As yet unmindful of the necessity for these precautions, he hurries, when the first generous freshet comes down, to the scene of his labours, and thrills with joyful pride as he recognises that, so far as can be detected through the watery turbulence, not a single boulder in his work has been dislodged. With the eye of faith and a glow of satisfaction he envisages shoals of grateful trout restfully poised in the shelter afforded by his dams from the wild buffeting of the surface current. By and by a stronger flood comes down, and still the dams hold. He has builded well. But as the flood runs off and the water clears a new fear is sprung on him. The pools are not as they were. Where he had left a depth of 2 feet of water just above the dams that depth has already diminished by 6 inches or more. He is seized by the horrible suspicion that his pools are silting up and that his labour has been in vain.

There is a good deal of truth in this suspicion. It is the natural thing, in the absence of counteracting influences, for a pool, formed by a dam which also acts as the spillway, to silt up. I have in my mind several instances of comparatively elaborate masonry dams, which have cost hundreds of pounds in building them across rapid rivers, where the pool has filled up to the height of the dam, so that the eventual result has been, instead of impounding the water, the raising of the bed of the stream immediately above the dam some 8 feet or

FIG. I.—THE POOL AS YOU MADE IT.

10 feet higher than the watercourse immediately below the dam. Obviously, the speed with which this silting up takes place varies with the character of the stream and the nature of the ground over which it travels. In the case of the typical mountain stream flowing over numerous stretches of coarse, easily-shifted gravel, and subject to, for its size, violent spates, the rate of silting up is very rapid. One wet winter may suffice to readjust the sectional contour of the pools to Nature's liking, after which the floods carry the superfluous silt further downstream.

Yet all is not lost—far from it. For one thing the dams have spread the water over a much greater area, which has an important bearing on a later problem, and as a compensation for shallowing the tail of the pool, it will be found that the floods, in tumbling over the dam above it, have scooped out great hollows in the stream-bed at its head. Natural forces have topsy-turvied your plan; your deeps are filled up and your shallows are made deep, as shown in the longitudinal sectional drawings.

Think of it! After all, Nature has but applied to your crude beginnings her own way of making a

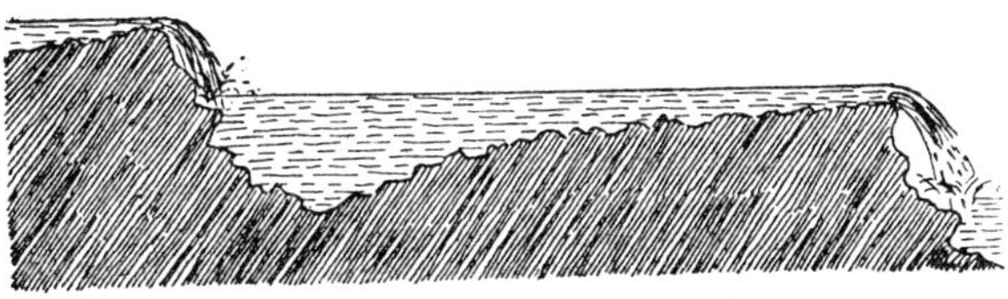

FIG. 2.—NATURE'S RE-ADJUSTMENT.

pool. Call to mind the typical natural pools on a mountain stream. All shoal towards the tail of the pool; the deeper water is at the head where the plunging current has dug out a cavity. Where there is a natural cascade or a plunging rapid the rule is that a pool forms immediately below it. Nature, with the assistance of her own cascades, scoops out a series of pools in the course of the stream. The lesson to be learnt is clear: Furnish her with several more cascades, carefully selecting sites where the scooping is easy, and let her scoop out a lot more. A gravel bottom she will promptly deal with, but it is useless for the present purpose to build a dam in such a position that the overflowing water falls on a huge slab of solid rock *in situ*—unless, of course, you have no foolish prejudices against waiting a few thousand years until the water has worn a basin of serviceable dimensions in the rock.

Here, again, there is need to look ahead, with a respectful recollection of the house that was built upon the sand, and of its fate when the rains descended and the floods came and beat upon it. If the gravel is very easily shifted, it may well be that the same force which scoops out the pool undermines your dam, should its base not be adequately protected on the down-stream side. Some considerable degree of such protection may be afforded by deeply embedded boulders immediately below the dam, and by stakes deeply driven into the bed of the stream, examples of both being shown in the last chapter.

The advantages to be derived from making your dams in quick succession should now be apparent. The long pool formed by a low dam which also acts as a spillway is almost, if not quite, impracticable on a mountain stream. Put your dams far apart, and you get between each pair a pool at one end where the rushing water comes in and a long stretch of raised stream-bed below it, after Nature has supplemented your efforts with her own. Put your dams fairly close together, and you get the terraced series of pools that bring comfort to the trout and joy to the angler.

# CHAPTER IV

## BOARD DAMS

In the work of creating desirable pools on a mountain stream the board dam, if not so picturesque as the rough boulder structure previously described, is simple and effective. Moreover, it possesses one great advantage over the boulder dam, insomuch as it offers certain facilities for dealing with the silting-up problem and maintaining, or renewing with little trouble, a serviceable depth of water throughout the length of the pool.

In its crude form the board dam is also one of the expedients of the moorland shepherds. I recall a favourite "wash dub" with which I am very familiar. Here the stream runs over a formation of solid limestone. At the foot of a tumbling rapid the water has worn out a large basin in the rock, and out of this pool the stream runs off over a flat slab of limestone. The pool thus formed is, in its natural state, too shallow for sheep-washing purposes. Therefore the shepherds throw across a temporary dam, on the flat immediately below the basin, by placing a couple of boards edgeways up, overlapping in the middle, and supported in that position against the weight of the current by a lump of rock in midstream and other pieces of rock at the ends. By this simple contrivance ten inches or so of additional depth are gained in the "dub," in which, and by these means, thousands of sheep have been annually washed from time immemorial.

During the washing season the boards are often left across the tail of this dub for weeks at a time, during which period the enlarged pool is well tenanted. I have seen trout rise in it within an hour of the finish of sheep washing, and from a vantage point afforded by the end of a wall bordering one side of the pool, have watched for hours the actions and manœuvres of its augmented population. It would be quite safe to say that the placing of the board across the stream certainly doubled the number of trout harboured in the pool, and I honestly think it trebled it.

In practically every upland stream flowing over solid rock you will find, varying in size of course, similar rock basins, many of which would be vastly improved by a few inches of additional depth. Peer down, with the light at the right angle, into one of the smallest of such rocky basins, and you are pretty certain to see, say, a couple of trout which have made it their resting place. Add 6 inches to the depth and you almost certainly find that the number of fish finding harbourage there has doubled. But that is not all. Those few extra inches of water, judiciously provided at the right places, may, by easing the path to the nuptial bed, immensely increase the value of the burn as a breeding stream.

Let me cite an example: On a certain hill

THE SHEPHERDS' BOARD DAM.

[*Facing page* 23.

stream, half a mile from its junction with the river, a little fall precipitates itself into a rocky basin. The spawning beds are all above the fall. At the spawning season a modest rush of water brings a little crowd of breeding fish up from the river intent upon family matters. But the head of water which has allured them up the stream to the foot of the fall is not sufficient to afford them passage over that obstacle. The little rocky pool becomes a congested area, out of which the natural instinct is continuously catapulting flashing forms, only to be washed back by the falling water. Not one trout in fifty succeeds in leaping the cascade. With a long-handled landing-net I have dipped scores out of the basin and lifted them over the fall into the water above. So closely packed in the basin have they been that I have often dipped out two, and occasionally three, with the net at once. Left to their own unaided efforts under such circumstances, the trout have the choice of waiting in the pool until a heavier freshet comes down, or dropping back to the river. Sometimes the heavier freshet materialises, and all is well; sometimes it does not—at least, not in time to be of service to the spawners—and then, as the novelists put it, when they are not quite sure of the logical consequence. . . . !

One alternative is that, heavy with spawn as they are, they fall back to the river, there shedding their ova, where it is promptly devoured, and not one embryo trout thus circumstanced comes to maturity.

With 6 inches of depth added to the basin any volume of water sufficient to bring the trout up to the foot of the fall would also be sufficient to give them an easy passage over it. At such spots a single board placed across the stream just below the basin would in many instances render the fall negotiable at any time. Ordinarily it may be somewhat difficult to find secure anchorage for the board. The ends, however, may generally be secured where they meet the bank with the exercise of a little ingenuity fitted to the circumstances of each case, while a midstream support can frequently be supplied by driving a spiked iron rod into a fracture in the rock, or it may even be worth while to drill the rock and cement in a short iron stanchion.

For the purpose of forming pools where the stream runs over a gravel bottom the task is a straightforward one. Having selected your site, you sink a narrow board across the stream and in the gravel bed to form a sill for your upper board to rest upon. The thickness of the boards may be anything from an inch upwards, as may be most convenient, but it must be borne in mind that the stouter the boards the fewer will be the stakes required to support them. Having supplied this narrow board for the sill, the upper edge of which should be raised a few inches above the normal bed of the stream, upright stakes should be driven in, close to it but in "staggered" order, above and below it. In the case of a narrow stream, three stakes will suffice—two below, dividing the length

THE REMOVABLE BOARD DAM.

of board roughly into three equal sections, and one stake above in midstream. Across wider becks and burns three stakes below and two above, as shown in the accompanying sketch, may be required. The upper board is then slipped, edgeways down, on to the sill between the two lines of stakes which hold it in position and support it—as far as the lower line of stakes is concerned—against the force of the current. The ends of the board may be wedged in position against the banks by boulders. An 11-inch board will back the water upstream for a considerable distance and make a fine pool.

The advantage of the board dam is now apparent. As the lower end of the pool silts up slide out the top board and lay it aside until the next freshet has come along and swept away the accumulated gravel, when, seeing that the sill is clear, you slip back your board, and, hey presto ! the same force which silted up your pool has cleared it.

It may be that the pool thus cleared is the top one of a series, in which case you have sent the silt down to assist in encumbering the pools below. However, that is merely one of the little items that make mountain stream preservation interesting, and by the entailment of a little labour relieve the dreary ennui attendant upon chronic perfection. By lifting the boards in succession—say, by having always one up, and five down in a series of six—the pools may be kept in a fairly clear condition.

In deciding whether you should put in a boulder or a board dam at a certain spot, you should use your own judgment, permitting yourself to be guided by the character of the stream and the availability of materials. Thus, on a length where the fall is rapid, where suitable boulders abound and Nature has anticipated your efforts to the extent of providing an irregular line of deeply-rooted boulders for the foundation of the dam, you will, no doubt, adopt the boulder system. Where, however, the fall is more gradual and the stream bed consists of shingle, you will probably find it more convenient to put in board dams with longer pools intervening.

I remember once sitting in solemn conclave, with two others—one of them the manager of a noted commercial fishery—on the fate of a charming private trout stream which had an assured volume of water from a corporation's compensation reservoir away up in the hills. That corporation was seeking powers to reduce the volume of the water in the stream by two-fifths, and one of the problems we had to consider was: How to maintain the efficiency of the stream as a trout fishery after the abstraction of such an alarming proportion of its water. After weighing the thing in our minds we compared notes. We had all arrived at the same conclusion—the solution of the problem lay in the introduction of board dams.

## CHAPTER V

### STAKE-AND-WATTLE AND BAG-WORK DAMS

To ensure a reasonable amount of satisfaction at the stocktaking of results, the quickly constructed dams hitherto described must be kept low. A "single-story" boulder dam may retain its efficiency and defy the winter floods for years if the boulders are securely locked in position. At the worst, if one boulder becomes dislodged another can usually be rolled into its place before the gap widens into an alarming threat of general dissolution. But to attempt anything in the way of an "upper story" is folly; for with the first energetic flood away goes the superstructure, and you have supplied yourself with another example of the downfall of overweening ambition. Yet there are points in your stream at which the construction of a dam giving a depth of 3 or 4 feet would be a tremendous benefit. In the case of one stream in which I was intimately interested, there was a normal flow of water of only a few inches in depth between deep banks. For a special purpose I wanted a depth of about 5 feet. This I got by having built a solid concrete wall across the stream from bank to bank. The stream did not come over this dam, but escaped by way of a sluice in the side of the pool. This was also built of concrete, and, with the silting problem ever present in my mind, I had the gap carried down to the full depth of the pool. The concrete sides of the gap were grooved so that the gap could be closed, or its height regulated, by sluice boards slipped down the grooves, one above the other. At any time which was thought desirable the whole of the sluice boards could be removed and the pool flushed out. I might add incidentally that the clerk to the local parish council, fired with a noble resolve to frustrate any deep-laid plot to deprive the community of its sacred water rights, assiduously watched the proceedings from a distance through a pair of field glasses.

Now, while this concrete dam proved highly satisfactory for the purpose for which it was intended, the cost of its construction was far too great to endow it with any recommendation for its emulation on the mountain stream. Still, it offered valuable suggestions. There are occasional spots in the course of the upland burn where a substantial dam of 3 or 4 feet in height would give a long and lovely pool for fly fishing. Moreover, on such stretches, there is often a depression in one bank or the other, halfway or two-thirds of the way down the pool, where, by raising the water 3 or 4 feet, a new outlet could easily be formed and the stream so diverted that it rejoined the natural bed at a point below the new dam. The adoption of this plan provides a good fly-fishing bit of water, gives the fish a free passage upstream, and increases the water-covered area, which has a material bearing on the commissariat department.

A substantial dam, well adapted to such an undertaking, can be, at a small cost if the labour is given by a faithful band of enthusiasts, thrown across the stream by combining the stake-and-wattle, sandbag and cement principles. Five-and-twenty years ago I saw weirs of this composite description put into a rapid river. So far they have withstood the violent floods of a quarter of a century, and to-day have every appearance of being able to defy the cataclysms of another quarter-century at least.

The materials you will require—and probably have to pay for—are a sufficiency of stout stakes, a few bags of Portland cement, a number of empty jute bags (sugar bags or potato bags) and a length of telegraph wire. The remainder of the materials consists of willow branches (which you will cut on the banks of the lower lengths of the stream, in many instances overhanging branches which you would in any case lop in order to give room for casting) and sand, which latter you will dig out of the bed of the stream and pass through a riddle in order to rid it of the coarser gravel.

In the accompanying sketch the principle of construction is plainly shown, the wattling being cut off halfway across the sketch to expose the bags of concrete lying between the two lines of stakes. Also, in order to exhibit clearly the method of interlacing the branches basketwise round the stakes, these branches have been given

COMBINED STAKE-AND-WATTLE AND BAG-WORK DAM.

a neatness and orderliness which they are not likely to attain in actual practice. For such a dam as we have now under consideration two rows of stakes should first be driven deeply into the bed of the stream, with a space of 18 inches between the two rows. Now we come to the wattling. Strip the willow branches of the minor shoots and twigs, and lace them in and out, alternately before and behind the stakes, until you have built up two rough basket walls to the height required. Of course, you have selected a time for the operation when the stream is running very small, and what water is coming down finds its way during this stage through the interstices in the wattling.

Next you have got to fill in the space between the two basket walls with what harbour engineers call "bag-work." In building up submerged foundations for piers and breakwaters it is usual, I believe, to use a "4 to 1" concrete, that is, four parts of fine, dry sand to one of cement. In the case of your dam a "1 : 4 : 7" mixture should serve and economise on cement; that is, one part cement, four parts dry fine sand, and seven parts small stones. The cement and sand should first be thoroughly mixed dry on a board platform with a shovel; then water is added and the mixture worked into a mortar, after which the small stones are dredged in and the whole is well turned and mixed with the shovel. Now you shovel the mixture into your bags and tie them securely round the neck. The bags must not be filled tightly. They must be left "sloppy" enough to permit the contents to sag readily and accommodate themselves to any irregularity of contour on which they rest. Drop a layer of these bags on the bottom of the wattled trench and press them into position against the bed of the stream by walking over them and treading them down. Repeat the process by adding more bags of the concrete until you have built up the bag-work into a compact mass to the height of the wattling.

And while you build you weave the thick wire in and out and through the combined structure, care being taken that it is not allowed to form any interstice by holding one bag from packing against its neighbour, criss-cross, round the stakes, anyhow and everyhow, so that it draws the component parts together tighter than the cords of love and binds the one to the other closer than a brother, preventing the rows of stakes being thrust apart by the bag-work.

Of course, the same style of dam may be put across the stream where there is no opportunity to make an outlet in the side of the pool and where the flow of water must come over the dam. But I confess that, from the angling point of view, I do not love either wire or wattle in a dam which is also a spillway in a small stream. Here, fishing upstream with a short line, your flies are frequently washed over the dam from one pool into the pool below. If there is one bit of submerged wire or miniature snag in the dam your tackle will fly to it like a homing bird to its nest and it will be the recording angel's busy day—on the debit side—as he enters up your uncharitable thoughts towards the man who put in that weir without planting over it a conspicuous notice board bearing the warning: " 'WARE WIRE!"

On the broader river the situation is different, for there you have more freedom of action and a wider choice of approach; but this touches a side of the general subject of improvement which is dealt with in later chapters concerned with the treatment of moorland rivers.

# CHAPTER VI

## THE FOOD PROBLEM ON THE MOUNTAIN STREAM

PROVIDED the stream empties into a river carrying an average head of trout, the man who tackles the job of mountain burn preservation need not worry about artificial stocking. The owner of the river fishery attends to that little problem for him. At first glance this sort of partnership may be thought to savour of the working formula of the bigger of the two boys who had clubbed up to buy a cigar: "I'll smoke and you can spit." In reality, however, the river man does get a *quid pro fario,* so to speak, for, even if a portion of his breeding fish do linger among the enhanced delights of the improved stream, the breeding conditions are materially bettered, easier access to the redds is assured, and, on the whole, better-conditioned fish and more of them drop down into the river.

With the fry that are hatched out in the stream and the spawners that come up at the back end, at one period or another of the year the burn swarms with fish. The question for the stream preserver is: Can he keep a fair share of them there? Somewhere I have heard or read of the village parson, meeting the wife of the village drunkard, and sighing reproachfully:

"I was pained to see John come rolling out of the 'Cow and Crumpet' again last night—drunk, as usual. Why can't you keep him away from the pub? Make his home more attractive for him and he'll stay there."

"Well, I dunno," came the weary reply. "I done all I could think of. I've pulled up the parlour carpet and sprinkled sand and spittoons on the floor instead, and I've had a barrel of beer put on chiffonnier, and I've pulled down 'God Bless our Home' off the wall and nailed up 'Bung's XXX Beers are Brightest and Best' in place of it, and if that don't make home more attractive for him what can poor mortal woman do more?"

An adequate food supply is one of the essentials in making home attractive to the trout. You have already provided the suitably furnished "parlours" by erecting dams. It may now seem to you that you have to begin to consider the problem of getting provender into them. Probably unconsciously, you have already taken a long stride in that direction. The ordinary methods advocated for increasing the natural food supply in lowland trout rivers will not serve when applied to upland becks. You cannot, with any degree of success, plant pond vegetation in mountain torrents; snails are swept away like marbles before a hostile broom. You cannot cultivate turnips on the rugged storm-swept heights, but the same heights will grow valuable mountain pasture if given the chance. Similarly, the mountain stream

THE SUITABLY FURNISHED PARLOURS.

[*Facing page* 36.

will produce its natural crop of food, and the wider its area can be extended the greater will be its yield.

"But," the reader may object, "all this does not increase the amount of food in the stream per square yard." Quite so; how that may be done is shown later. But if you can double or treble the amount of food per longitudinal yard of the stream it is well worth while. Again, take as an illustration the mountain pasture. If you limit the range of a flock of sheep to a certain acreage, they may have a fight to drag on a half-starved existence. Double the acreage of their pasturage, and they may do very well indeed. The same applies to the trout of the mountain stream.

Earnest labourers in the trout food production field have covered reams upon reams of paper with fervent appeals and exhortations and arguments, excellent so far as they have gone, but it appears to me that they have always been so obsessed by the intensive culture principle that they have invariably overlooked one of the most important points affecting more especially mountain streams and moorland rivers. On these streams aquatic insects, in various stages of their cycle of existence, form the chief food of the trout—and you cannot raise a crop of aquatic insects on dry land. For a considerable portion of the year a large part of the stream bed is simply a waste of dry gravel—put out of action as the breeding ground for aquatic larvæ.

Perhaps the best well-known fact I can adduce

by way of illustration is the case of the gnat, or mosquito, nuisance. The mosquito is an aquatic insect, breeding in shallow, stagnant water. Drain your ditches and ponds, the mosquito disappears; increase the area of your stagnant pools, and the mosquito multiplies proportionately—does it himself without putting you to the trouble of importing and laying down eggs from a distance. The bulk of the aquatic insects of the stream have a life history resembling in rough outline that of the mosquito. What follows then? Double the width of your stream, and, if you do not with mathematic exactitude double its crop of aquatic insects, you increase it enormously.

Take the case of the Stone-fly, one of the most valuable items in the larder of the trout streams running over rock and gravel. It is nourishing and sustaining; also it constitutes the most generous mouthful aquatic entomology can insert in the menu. One single Stone-fly, or "Creeper," will go as far to fill a long-felt want as fifty or a hundred "Cusses." The class of stream we have had under consideration is the natural home of the Stone-fly. The number of "Creepers"—the nymphs or larvæ of the Stone-fly—you stir into evidence while working in the stream bed during the making of your boulder dams will probably come to you as a surprise. But the stream is naturally narrow, while it is bordered by an expanse of dry unproductive gravel. Take the dry area into cultivation by covering it with water—as you already have done to a great extent by making your pools

POOLS THAT HAVE DOUBLED THE WIDTH OF THE STREAM.

[*Facing page* 39.

—and you double your previous Stone-fly crop on that section.

The question may naturally arise: Will nature kindly sow the good seed in additional quantities on the new ground, or does she already distribute all the available supply on whatever area of water happens to be there? Analogically, the mosquito illustration supplies the answer. Directly, all the evidence that years of acquaintance with the insect have tendered makes me absolutely certain of the soundness of my contention. Nature is prodigal in her waste of materials in her reproduction scheme. If all the eggs produced by the fertile female Stone-flies were deposited in the streams and hatched out the waterways would be completely choked with a mass of wriggling larvæ. But the gravid female is a wanderer on the face of the earth, and not all the policemen on point duty could direct the erratic reproductive Stone-fly traffic to the narrow waters of its origin. Possibly Nature has a definite purpose in endowing the female Stone-fly with the roaming habit at the risk of her dumping her embryo progeny more or less promiscuously on the face of creation. It may be Nature's system of colonisation—her scheme for spreading the race. If only some of the females wander far enough they may come upon new water in which to establish the family.

If that is Nature's object, success can only be attained at the enormous expense of millions of wasted eggs. You can encounter gravid females, with great clusters of eggs ready for deposition, long distances from the water and apparently lost to all sense of direction. I have seen thousands crawling about white, dusty, limestone roads in the vicinity of moorland rivers when the hatch has been at its height, only to be squashed under the feet or wheels of passing traffic, and I have often wondered if, in some short-sighted fashion, the creatures do not in such cases mistake the glare of the white roads for the glitter of water.

Anyhow, whatever justification may lie in these speculations, the broad fact remains that in the later days of May and the early part of June, prodigious quantities of Stone-fly eggs are wasted which, with a bigger reliable area of water, would fall upon productive ground. By "reliable" I mean an area covered with water *which will remain covered.* This is assured by your dams. Otherwise it might occur that the usually dry patch of gravel bed is temporarily submerged by a freshet at the time eggs are deposited upon it. The freshet runs off, probably to be succeeded by a spell of drought, during which a considerable share of a potential generation of fly on that section of the stream perishes and is lost to the trout before it has attained any food value. Not only that, but as the floodlet waxes free-moving larvæ spread out over the temporarily covered gravel, and many of these, too, are left stranded when the freshet wanes.

I may seem to have dwelt at undue length on the subject of increased water area, and it will be seen that in later chapters I have returned to the question as it relates to the food problem of the moorland river. I have done so because, while to my mind it is the first thing that calls for attention in the commissariat of such waters, it is a point which is generally overlooked.

# CHAPTER VII

## THE LARDER OF THE MOUNTAIN STREAM FURTHER CONSIDERED

THE story is told of a hospitably minded Highland laird who was given to inviting acquaintances casually to drop in on him for a day or two's fishing any time they happened to be in his neighbourhood, with the consequence that, the river being in fine ply, a dozen of these casuals dropped in simultaneously.

"What wull I do, Dugald?" exclaimed the laird despairingly to his trusty henchman. "Here's a dizzen of 'em, an' we've only spare beds for five!"

"Dinna fash yersel owre sic a wee matter as that," promptly returned Dugald. "Gin ye gie 'em eneugh whusky the chiels 'll find beds for their ainsels."

I quote this story because, antithetically, it illustrates so aptly the position of many of your entomological guests on the mountain stream. Reversing the order of Dugald's philosophy, give them beds and the insect chiels will find the means of satisfying their gustatory requirements their ainsels. Take the case of the *Ephemeridæ*. Your dams, by increasing the water-covered area, have found additional water for the swimming nymph, such as the Iron Blue; the flat nymph, such as the

March Brown ; and the crawling nymph, such as the Blue-winged Olive, to disport themselves in. But they have done more ; they have provided a habitat for the digging nymph, of which the Green Drake, or Mayfly, is the most important example.

I believe that, more than on anything else, the prevalence, or scarcity, of the Mayfly depends upon the prevalence, or scarcity, of suitable housing accommodation during the nymph stage of its cycle of existence. You may live on love—and a few other necessaries—in a cottage ; but the cottage is essential to the success of the scheme. I am under the impression that a—more or less—general notion obtains that the Mayfly rise is, for all practical purposes, confined to the lowland rivers. This is a delusion. The Drake has its home on hundreds of upland streams at points where Nature has deposited a bed of fine sandy detritus for the nymphs to burrow in, and such spots are to be found at the tails of natural pools a thousand feet and more above sea level. On many such pools you can see the Green Drakes hatching out merrily in June. Your artificial pools in the same streams will produce like phenomena, and thereby increase the food supply materially. This was, in a mild way, sensationally impressed upon me by a visit several years ago to the stream mentioned in the second chapter.

Following its course downstream, you come at last to a remote moorland farm which was at one time also a tiny mill, and immediately above the farmstead the stream has been dammed to form a miniature mill pool. On the occasion of this particular visit I had not a thought of Mayfly in my head on leaving home, and, consequently, on arriving at the stream, was amazed to find myself in the thick of one of the densest clouds of Drakes I ever saw in the north of England. Along the stream the hatch extended, fish rising to the fly in all the natural pools, but by far the most congested area was the immediate neighbourhood of the artificial mill pool, where the dam had collected a bed of fine detritus.

Incidentally, I might add that I had no single Mayfly in my fly-book, but I happened to have a March Brown, which I mounted, and on that killed over twenty trout before either the fish went off or the March Brown wore itself out in honourable tatters.

Caddis-worms appear to do well in the mountain stream and seem to have the quality of resisting the strong floods. Perhaps, in the case of some varieties, their armour of tiny stones ballasts them helpfully. I have more to say about them in regard to moorland rivers. In the upland becks we may leave them to take care of themselves, with few words. Should any attempt at fly cultivation, beyond the improvement of the conditions favourable to their natural propagation, be attempted on the mountain stream, the first care of the culturist should be to increase the stock of those varieties indigenous to the stream. Should he take a step further by attempting to introduce new varieties, he should assure himself

THE MOORLAND FARMSTEAD WITH THE MAY-FLY RISE AT ITS BACK-DOOR.

[Facing page 44.

that his selection is limited to those species which are known to breed freely in waters similar to those at his disposal. It is as foolish, for instance, to attempt to stock the rackety upland burn with lake flies as it would be to house the ferret in the canary cage and attempt to cajole him into living on bird-seed. Many of the smaller caddis-flies are common on the upland streams, such as the Siver-horns, and, generally speaking, the caddis-worms found in the moorland rivers of the district may, with reasonable hope of success, be gathered and transplanted in the mountain stream in the pools formed by the dams.

Many of the *Ephemeridæ* inhabit the mountain stream, and in the construction of the dams their opportunities for self-stocking have been materially increased. Lack of success and profusion of failure which have usually attended efforts in that direction do not encourage any attempt at trans-plantation with the duns. With the Stone-fly family, there is much greater hope of gratifying results. Should, by any possibility, the large Stone-fly be absent from, or scarce on, your mountain stream, eggs of that insect should be procured and laid down, while the Needles and the Willow-fly, belonging to the same family, are very valuable trout foods which may with advantage be cultivated.

Leaving the flies, the freshwater shrimp plays a leading part in stocking the larder. The creature is quite at home in the upland waters. I have many a time seen it frolicking in the roadside horse troughs beside the tracks over the mountain passes, and where watercress can be cultivated the supply of this creature can be hugely increased, but, as I have already said, you cannot grow aquatic plants—except the moss that clings to the boulders—with any degree of success in the main stream. Indeed, the planting of even marginal plants is, for the most part, discouraging. Willow slips may be stuck in the banks, where they take root and do good service, but, beyond them, one's botanical efforts might, in the case of the mountain stream, be largely confined to watercress (*Nasturtium officinale*) and marsh marigold (*Caltha palustris*), where opportunities offer themselves.

As the old salt, when asked what would be his three wishes if he could have such gratified, replied, " Well, I'd have all the baccy in the world ; then I'd have all the rum in the world ; and then—then—well, then I think I'd have a bit more baccy " ; so would I say, with regard to the planting of mountain streams, " I'd have all the watercress I could introduce ; then I'd have all the marsh marigolds I could work into the scheme ; and then—well, I think I'd have more watercress." There is many a patch of wet ground near the stream on which watercress would grow and the marsh marigold would flourish like the proverbial one o'clock. Say a score of yards back from the beck, water from some hidden source seeps out from a miniature bluff in the hillside, and gently oozes through spongy sod until it trickles into the stream. Strip the sod, make a watercress and marsh marigold bed, and let the water carry with it into the stream a useful commissariat contribution. Or it may be that at some point a little by-stream breaks away to rejoin the main bed 50 or 100 yards lower down, in which case this might be utilised, with some provision against flooding in times of spate. Another means of furnishing watercress beds will be found included in a later chapter.

## CHAPTER VIII

### FROGS AND SNAILS

In our nursery days we were assured on classic authority that little boys were made of

> " Frogs and snails
> And tadpoles' tails."

and the same dietary ingredients go to the making of plump, healthy trout when they can be supplied in adequate quantities. In the case of small trout, of course, the frogs should be in the infantile stage—just emerging from the tadpole phase—and the tadpoles' tails need not be divorced from their heads. All are welcome. The trout take them voraciously when they get the chance. Give them the chance. How ? Here and there, within a few yards of the stream, are depressions in the ground filled with water which may, perhaps, have percolated through the gravel. In spring these little pools are sometimes well-nigh choked with jelliferous masses of frog spawn which by and by develop into tiny wrigglers, and, in a short time, the pools are alive with crowds of lively tadpoles. Left to themselves, their final metamorphosis is completed, they hop ashore as little frogs, and are lost as potential trout food. Where such pools exist each should be considered with regard to its special features, when the enthusiast will usually be able to devise some practical method of dealing with it on its merits, even if, in the case of a very small pool, he has to go the length of netting the tadpoles out with a fine-meshed net and releasing them in the main stream.

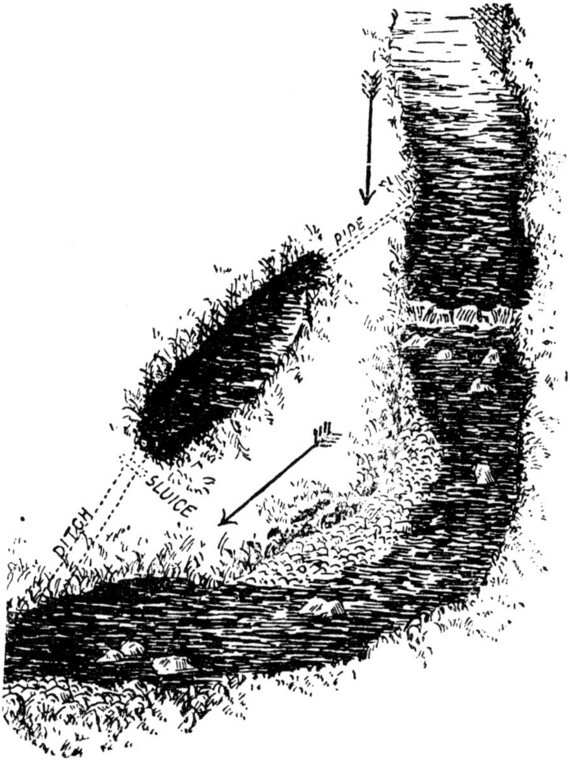


AN ADMIRABLE FOOD FACTORY.

In one instance in which I was some years ago interested, Nature could not well have planned the situation better. A pool extremely popular with connubial frogs occupied the angle formed by a quick bend in the stream, as shown in the sketch. Each spring that pool was black with a wriggling mass of tadpoles. I yearned mightily to put a length of 4-inch pipe through the bank from the stream above to the head of the pool (with a plug in the stream end of the pipe), scratch out a narrow communicating ditch between the tail of the pool and the stream below, then, just as the tadpoles had reached their fattest, juicest condition, pull out the plug in the stream and give the trout a club feast by flushing out the whole comestible caboodle into the beck. But, for some reason, the owner was not entirely sympathetic, and so was lost a chance of establishing an admirable food factory, which might have been utilised for the production of varieties of delicacies other than tadpoles. It is possible that snails would have done well there—I have heard of them being cultivated in such side ponds—and cajoled in instalments down the ditch into the stream at times when the conditions offered a prospect of the trout getting them before they were swept away. In the sketch I have shown the ditch and flushing pipe by dotted lines, and where, at the head of the ditch, some form of rough and ready sluice might have been put in to complete the control over the outfit.

I have said that in the riotous mountain stream snails, which, where available, form one of the most valuable trout foods, are bowled away like marbles, and in such streams they are seldom found. Experiments in snail culture and transplantation during recent years have led to one conclusion—that there is little hope of marked success in attempting to introduce freshwater snails into water in which they do not occur naturally. Their absence from the majority of our mountain streams is probably due to one or other, or both, of two causes, viz., the violence of the current, or the absence of calcareous matter in the water from peaty sources.

In the less boisterous tributaries, impregnated with chalk or limestone, and especially if such burns are provided here and there with slacks and deeps, there is no reason why snails should not be found in, at any rate, small quantities. Where that is the case, there is every encouragement to try snail culture—concerning which information will be found in Chapter XV.—with every hope of gratifying results. That such exceptional streams do exist, I know, although I have had no personal experience with them, and, as an example, I quote the following note from a correspondent which appeared in the *Field* while the bulk of the matter forming this book was appearing serially in that journal :—

"About twelve months ago there appeared a paragraph (from Dr. J. C. Mottram) explaining the advantages to be derived in a fishing stream from the culture of freshwater snails. After getting into communication with the writer, I received clear explanatory directions and sketches, which were adopted upon the small stream upon which six rods hold the sporting rights. I am now most pleased to be able to say that investigations proved that the stream in question held a few freshwater snails, and more of them were collected from adjacent waters, and sowed down in small runners and backwaters where the trout were unlikely to find them. The results have been most encouraging. We have, to-day, literally thousands, and they are spreading downstream in a most satisfactory manner. It is felt that the introduction of snails forms a most useful addition to the trout feed of the stream, and though the water is merely a brook, pounders are by no means uncommon, and fish up to nearly 1½ lb. are occasionally taken, which must be admitted is quite good for North-country water."

## CHAPTER IX

### THE MOUNTAIN STREAM—WITH VARIATIONS

PUT summarily, and Drydenised, what we look for in our ideal youthful watercourse of the rugged uplands is :

A stream so various that it seems to be
Not one, but all streamkind's epitome ;
Swift in the changing keynote of its song,
Is everything by turns, and nothing long ;
And in its course a-down the mountain side
Is cascade, eddy, rapid, pool and glide.

When the reader has rectified his first erroneous impression that the over-page sketch is a surgical chart of the alimentary canal and the digestive system, the figures may assist in showing him at a glance how something substantial may be accomplished towards the attainment of the ideal by the adoption of means already described, supplemented by other aids.

Here and there, sooner or later, our typical mountain stream strikes a length of valley, where the trough of the ravine has been robbed of its V-shaped acuteness by an age-long collection of débris covered with rugged pasture, and over this coarse gravel the stream flows in one long rippling shallow, its devil-may-care boisterousness abandoned for the time being. It provides, perhaps,

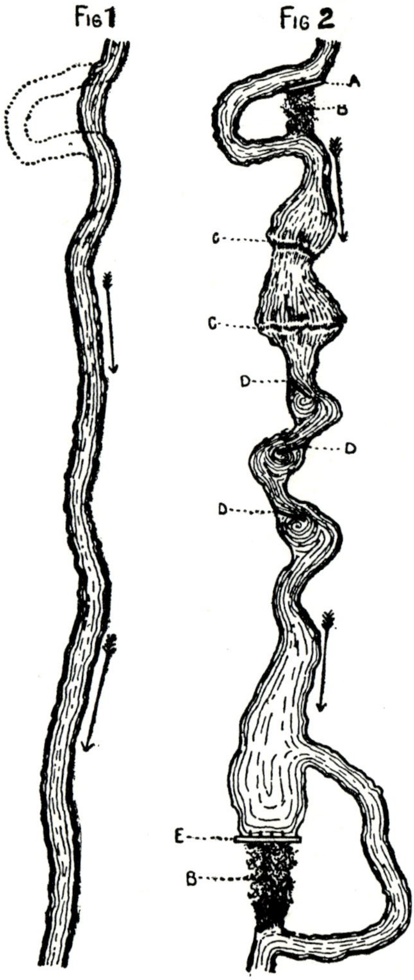


BEFORE AND AFTER TREATMENT.

good feeding ground for the trout hunting the Creeper or other larvæ; but it possesses no deeps—no restful lies in times of stress. What can be done with it to remedy its defects? A good deal. Dam it back—put kinks into it—tie it up into knots—shuttlecock the current backwards and forwards from bank to bank. It is the crooked stream that constructs the cosy corners.

Fig. 1 represents such a length of stream—straightened out a little to meet the exigencies of the printed page—which, with little variation, dances decorously along, ankle-deep, over the shining shingle with little more break in its character than is provided by an occasional small outstanding boulder. The dotted loop at the head of the figure shows the original course of the stream at this point where repeated floods have formed a cut-off, so that now only a little water goes round the bend on the occasions of strong floods.

Fig. 2 shows what might be done with such a length of stream, consideration of space again crowding the different features close together. At the point, A, close the breach in the original bank with a weir—probably stake-and-wattle and bag-work will best serve—of sufficient strength and height to throw the stream, even in time of ordinary flood, back into its former course, and in the channel of the cut-off plant a bed of watercress and marsh marigolds, B, feeding it by means of three or four 1-inch iron pipes inserted obliquely through the dam so that their upper ends incline a little downstream, thus avoiding risk of being choked by stream-borne gravel in a freshet. The food produced by this bed will find its way naturally into the main stream, and, if a suitable depth of water is maintained, the shelter of the aquatic jungle will afford sanctuary and sustenance to the fry.

C C are low dams. On such a length of stream as suggested in the foregoing where boulders are scarce, board dams, whose merits have been dealt with in a former chapter, will be found most serviceable. The series, of course, may be continued to any number which may be thought desirable.

D D D are low groins of stake and wattle, set at an angle to throw the current against the opposite bank. They may be very simple of construction—just four or five stakes driven well down into the gravel, and interlaced, basketwise, with willow branches. They need not be more than 18 inches or 2 feet high above ground, according to the size of the stream. Having put in your groins, you wait with all the fervency of the child who opens his mouth and shuts his eyes to see what will happen next. Precisely what Nature will do when the freshets come along if you give her a lead like this it is impossible to foretell. The only thing you can with certainty depend upon is that she will do it all right, and it will be something interesting. Most probably each succeeding freshet will assist in piling up a bank of gravel on the upstream side of the groin until the wattle is buried on that side to average flood height; and the heavy current, flung violently across its original bed, will eat out the opposite bank until a miniature cliff appears, with the water deepening to a depth of 2 feet or 3 feet at its base. Behind the groins, *i.e.*, on the downstream side of them—will most likely be left fairly deep slack water, with, somewhere, and probably in the lee of the outward end of the groin, a deep eddy.

The why and wherefore of many of Nature's freaks in scooping out holes in the gravel beds of rapid streams are beyond the comprehension of the man of merely average intelligence, but it is certain such a series of groins as is sketched in Fig. 2 will transform a piece of commonplace, shallow, babbling brook into a fascinating bit of tricky and trouty water, offering lurking places to the fish and problems to the fisherman. It is, too, well within the range of possibility that some of the willow branches will take root in the gravel piled against the wattle, resulting in bushes overhanging the stream and, being energetic factors in the production of insect life, they will drop their showers of blessings where they are very acceptable.

At E is introduced a stout stake-and-wattle and bag-work dam to form a long, comparatively deep pool, from which, by raising the surface height, the water will find a new outlet at the side and cut itself a fresh channel, rejoining the old course at a lower point. Here, again, is a capital opportunity for establishing a watercress-cum-marsh-marigold food factory, B, for the mass production of freshwater shrimps as its staple output, with valuable insectorial side-lines. The bed, of course, is formed in the old channel between the dam and the point where the diverted stream comes into the original course, and, as in the former case, it is fed through 1-inch pipes, built in the dam, from the pool above. In this instance, the pipes may be laid at right angles with the line of the dam, for at this point there will be no current butting at the dam, and there is little likelihood of their choking up. In such an unlooked-for event it is an easy matter to clear the pipes of obstructions by thrusting a rod through them from below.

Exactly what will happen to this pool is also something of a speculation, depending on the contour of its banks, the strength of the stream when in flood, and the tenacity of the bank through which the diversion has been made. It may be that you will have to do a bit of fighting against too rapid erosion, and the consequent lowering of the water in the pool at the breach in the bank, but this little difficulty can be left until it arises, when a piece of weiring and a scrap of ingenuity will apply the remedy.

As I have said before, one cannot prescribe in detail for the idiosyncrasies of the unknown stream ; each case presenting at some point or other individual peculiarities which can only be treated to the best advantage after consideration of their special features ; but I think the suggestions I have made in this and the preceding chapters will be helpful in determining the class of treatment to be applied. To my way of thinking, the chief objects to keep in mind are : the formation—with the aid of what natural forces can be enlisted—by dams and groins of pools, holes and eddies, increasing by the same means and by diversions of the channel, the permanently submerged area, thus enlarging the natural food-producing capacity, and the planting of watercress and marsh marigold wherever possible, also the planting of willows on the banks. Where such advantages as I have indicated exist, whether they have been introduced naturally or artificially, the fish average far better than on the less fortunate upland burns. On one stream I have often fished the trout were a very level lot, running about four to the pound, and, while engaged in the preparation of this very chapter, a man has been in to tell me of a catch of seventeen he made on a certain mountain stream I know very well, in which he had one fish well over 1 lb., and several half-pounders.

Among the foes to mountain stream preservation must certainly be classed the heron. The graceful scalliwag is an earnest admirer of your work in fashioning fishing pools. As you leave your labour on the dams you may observe him planing down from the heavens to commence his. The dipper haunts your stream persistently. "Nature lovers" splash the whitewash brush vigorously over his bill of fare; they paint him in seductive tints as the faithful friend of the angler, gobbling up only the insectivorous enemies of the infantile fry, and, at any rate, they plead in conclusion he should be "given the benefit of the doubt." Now, I do not want to be dogmatic, nor do I wish to condemn the innocent, but when I am asked to believe that a dipper works so conscientiously and discriminately that he will joyously pounce on the larvæ, say, of the murderous *Dytiscus marginalis* and virtuously ignore a tempting and tender alevin my powers of credence are taxed to something a little over twenty shillings in the pound. I would not withhold from any deserving creature the "benefit of the doubt," but, somehow, when the case is that of the dipper foraging on the spawning beds of a trout stream, I have a vague, intuitive feeling, that the safety of the fry would be better assured if the dipper were shot first and given the benefit of the doubt afterwards.

The black-headed gull, which comes inland to breed on the high moorlands, is an inveterate fish poacher, and, in the words of an eminent apologist for the bird, "Makes havoc among the little fishes" when he gets the chance. He may be a good friend to the ploughman, but he is a curse to the preserver of upland streams and tarns. Some years ago a moorland tarn, in the close vicinity of a favoured nesting ground for black-headed gulls, was stocked with trout. The tarn was not far from my home at the time, and I knew it well. The imported trout did not appear to be doing well, and suspicion fell upon the gulls. One day the keeper, approaching while the gulls were busy on the tarn, fired among them. Up rose the birds in a flock, and in their fright disgorged a shower of over forty young trout!

Another natural enemy is the moorland farmer's

small boy. Like the mathematical point, what he does not know about "guddling," "groping," or "tickling" trout has neither length nor breadth nor thickness. His knowledge is diabolically complete and severely practical. Yet it is well to wink as far as possible at his depredations for the sake of keeping on good terms with the moorland farmer. In this world you never know whose wheelbarrow you may want to borrow.

## CHAPTER X

### ARTIFICIAL TARNS

OPERATING on the grand scale, wealthy municipalities have throughout our uplands improved the mountain streams, with their, comparatively, modest capacity, into vast trout fisheries, and, of course, have incidentally furnished far-away city-pent dwellers with domestic water. In the moorland valleys of the Pennines we have thousands of acres of sporting water yielding first-rate fishing, over ground where once the sheep nibbled and the curlew nested. Some of these corporation reservoirs are accessible to the angler by ticket procured from the municipal authorities ; in other cases the fishing rights are let to angling clubs, the members of which preserve and, if necessary, stock. When fed by a few streams offering good spawning facilities, many of these artificial lakes are recuperative, from the census point of view, to a wonderful extent. But whether naturally or artificially stocked the fish in them rapidly accommodate themselves to their loch-like environment and thrive exceedingly. Compared with native trout of the burn that once threaded its way down the valley, the average weight of the fish in a typical large reservoir will increase threefold or fourfold with plenty of pounders available to provide a special treat at intervals. We have notable instances of the possibilities, in the way of big trout, of impounded waters in the annals of Blagdon and Lake Vyrnwy. The records of the early days of the former were marked with numerous entries of trout of 6 lb., 7 lb. and 8 lb. Lake Vyrnwy was completed in the early 'nineties, and, in 1899, one angler with fly caught a basket of forty-three trout whose total weight was 37½ lb. The most sensational capture of trout that ever came within my notice was that made by a well-known Yorkshire angler on one of the Leeds Corporation's big reservoirs in the valley of the Washburn, a typical upland beck, a few years after its completion. With his own rod in one day he filled four fair-sized hampers with trout, the total weight of which was estimated at 6 stones at the least, and the catch included fish up to 2½ lb.

While these huge municipal schemes may logically be claimed to effect big improvements in the sport-yielding capacity of our upland waters, they would remain outside the scope of these chapters, were it not that they may often be copied in miniature at very little cost compared with the advantages which can be secured. Frequently a very small dam built across an insignificant burn will result in the flooding of several acres to a depth commensurate with the holding of a good stock of trout. It can be done ; it has been done.

The late Colonel Peter Ormrod once told me of a capital trout lake of some 20 acres he made by impounding the waters of a small mountain stream, the result being a fine sporting fishery. He wanted me to go over to see it, but, unfortunately, the war intervened, and I never visited it. On the slopes of Ingleborough, in the private grounds of the Farrers, a beautiful tarn of some 7 acres has been formed by throwing a short dam across a narrow ravine and impounding the waters of the Clapham Beck. There has been no planting of food-producing aquatic or marginal plants, the tarn does not seem to have been cultivated to any extent as a fishery, and yet the trout in it average about $\frac{1}{2}$ lb. in weight, while those in the beck will probably run about six to the pound.

Within ten miles, as the crow flies, of the Ingleborough Lake, is a sheet of water not unknown to literary, geological, artistic, and angling fame—Malham Tarn. Viewed under the glittering radiance of a summer sky that pours down molten sunshine on its sparkling surface, it is a magnificent liquid jewel worn on the bosom of the everlasting hills of the Craven Highlands. Situated on the rugged fells that go scrambling and climbing upward to the summit heights of Pennyghent, a few miles further north, the tarn is as high above sea level as the Dead Sea is below it—1,200 feet. On the south and east rough mountain pastures creep to the very margin of the lake, and Highland cattle, knee deep in the laughing ripples, complete the picture when seen at its best. On the west the tarn is bounded by grim peat hags, crowned in summer by a glorious blaze of purple heather; while the northern shore sweeps swiftly upward through pine woods to the white, cliff-like scars above, and among the trees are the sloping lawns, gay with beds of Alpine blooms, of Malham Tarn House, for more than half a century the country house of the late Mr. Walter Morrison, and now in the possession of his nephew, Major J. A. Morrison —an exceedingly comfortable and beautiful lodge in an extremely attractive wilderness.

Malham Tarn is full of fish. A generation or so ago it was full of trout and sensational baskets were made at frequent intervals ; but the late Mr. Morrison, charming companion and generous friend, held rather peculiar views on fish preservation, and the perch were allowed to get the upper hand, with the consequence that the tarn now swarms with them, while the trout have been reduced greatly in numbers ; though, individually, they attain a big weight, a couple of splendid specimens in the neighbourhood of 6 lb.—one just over, one just below—having been landed last season. Inshore, baskets of hundreds of perch have been made, the majority, of course, running on the small side, but in the wider waters rove gallant companies of hard-fighting warriors of heavier calibre.

On the last occasion on which I visited Mr. Walter Morrison at the tarn I went afloat for the special purpose of attempting the capture of one of the *Salmo* sockdollagers, the subsequent post-mortem on which was to prove some disputed point with regard to his diet. The Imp of the Perverse, who claims the angler as his special *protégé*, accompanied me in the boat. I caught no trout. True, I had a long and exciting battle with one of the leviathans, but that is another story. Still, fishing with an Illingworth light-casting reel and natural minnow, I had the briskest bit of perch fishing I ever enjoyed; for in less than an hour I filled a good-sized pannier with fine fish running from $\frac{3}{4}$ lb. to $1\frac{1}{4}$ lb., and, I understand, the keeper talks about it to this day.

Incidentally, I read the riddle of the tarn thus: The trout fry are devoured in great numbers by the perch. A few run the gauntlet successfully until they have reached a size which secures them from the jaws of the enemy, after which they apply themselves to the consolatory task of "getting a bit of their own back" by feeding largely on the millions of perch fry. In the endless struggle the perch have gained the ascendancy by reason of their enormous reproductive powers, the female perch spawning 50 or a 100 ova to the female trout's one.

Where such a set of circumstances prevail, drastic measures for the suppression of the perch should be undertaken and vigorously pursued if the trout fishing is to be preserved, and the spawning season in the early spring is the time for the crusade. From March to May, according to locality, the perch spawn in the shallow waters, entwining their ova in long ribbands among the reeds, piles and such submerged objects as are handy. Only a few days elapse before the fry hatch out, but during those few days ova containing millions of perch in embryo may easily be collected with the landing-net and destroyed, while swans and ducks will eagerly lend their assistance in the work of devastation.

So far I have tried to convey the impression that here, undoubtedly, is a fine piece of water abounding in natural food sufficient to the upkeep of a big head of fish in the pink of fighting fettle and culinary condition. What brings it within the province of this chapter and renders it an object lesson of especial interest to all contemplating the formation of a new tarn or the expansion of an old one, is the fact that it is largely artificial. The Malham Tarn of to-day covers 160 acres, and is nowhere more than 14 feet deep. Originally it was a natural pool formed by the accumulated waters of sundry springs and mountain streamlets collecting in a shallow basin in the Silurian rocks, and its greatest depth was 9 feet. At some time, however, a short, low embankment was thrown across the natural outlet, which, comparatively, trifling undertaking raised the height of the water 5 feet, and enormously increased the acreage, bringing the tarn to its present proportions, every yard of which is fish-holding or fish-feeding water.

The value of such an expansion, achieved at small cost, is corroboratively reflected, inversely, by the case of Loch Leven when, a hundred years ago, a drainage scheme which had been put into operation diminished the acreage by one-fourth and lowered the water some 4 feet or 5 feet. This difference was proportionately far less in extent than that which must have been effected in the enlargement of Malham Tarn, and yet the fishing was so disastrously affected that it was estimated it had depreciated one-third in value. There are many small tarns the fishing value of which could be doubled, or more than doubled, by raising the water a foot or two, and there are many instances of such tarns in which such raising of the water level could be brought about by simple and inexpensive means.

Of entirely artificial tarns made on mountain streams the latest example within my knowledge is Yew Tree Tarn, near Coniston, the construction of which, with details that may be helpful to those who may wish to do likewise, is dealt with in the next chapter.

# CHAPTER XI

## MAKING A TARN

In harmony with the eternal quest after the greatest results with the expenditure of the least amount of labour, the prospective tarn maker should carefully select a site which will, other things being equal, give him the greatest water acreage with the least barrage. The equilateral triangular tarn, with one of its sides composed of a long and costly dam, is out of the economic question. What the practical enthusiast sighs for is a site with a "bottle-neck," the weiring of which—with material procurable on or near the spot—will bottle up a proportionately huge area of water. Such a site is not infrequently encountered in following the course of a mountain stream.

Here and there the roystering beck comes to a spot where the opposing hillsides, instead of closely knuckling into one another, stand back a little to form a natural basin before closing in once more. Along the floor of this basin the stream loiters meanderingly for a spell, then escapes by the gap at the foot of the hollow to resume its rough-and-tumble flight to the river. Here is the ideal site for an artificial tarn. Dam the gap and you flood the basin, while the mountain sides and the torrent bed provide stone for the masonry of your dam.

Such a sporting lake, known as Yew Tree Tarn, was constructed especially for trout fishing purposes three or four years ago by Mr. James Marshall, of Coniston. It is a capital example of the quick transformation of a length of negligible mountain stream into a fine fishery run on high-class lines, and as, at the time of writing, the

FIG. I.—THE PRINCIPLE OF THE DAM.

latest object lesson of its kind it deserves more than passing mention. It is already a familiar sight to Lakeland tourists, for the road from Coniston to Ambleside, through the rugged scenery of beautiful Yewdale, skirts it on the east side close to the famous picnicking spot, Glen Mary. From the west side, through groves of fir and tangles of scrub, the fells sweep swiftly up in their rugged grandeur to the lofty skyline.

The tarn covers 7 acres, and this sheet of water

is held up by a dam whose total length is 40 yards and whose height is 13 feet above the original stream bed. For security's sake the proprietor went down 6 feet below the stream bed for the foundation of his weir, so that from base to summit the dam is 19 feet. With a better bottom a good deal of labour in this respect would have been

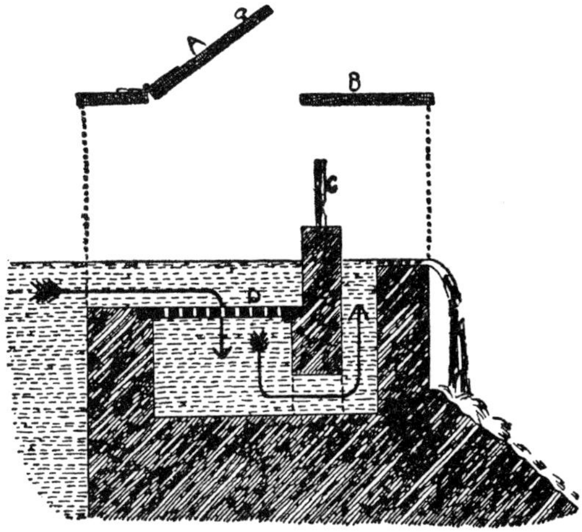


FIG. 2.—THE PRINCIPLE OF THE OUTLET.

saved. As it was, the local contractors shied at the job. They shook their heads gloomily and murmured, "Nea bottom." This did not satisfy the enterprising proprietor. Mr. Marshall and his head keeper (who had been for many years with Mr. Armistead, of the Solway Fishery) buckled to, with the assistance of a couple of labourers, built the dam themselves, and a capital job they made of it. During the three years it has been in existence it has been subjected to the violence of a plethora of severe storms and floods that have tested it thoroughly, and if such a dam will come through successfully on what the local pessimists described as "Nea bottom," then surely, on obviously sound ground, it can be commended as likely to defy the ages.

Fig. 1 shows in correct proportion the principle of the dam. The face of it is a solid wall of masonry and concrete, 2 feet thick, and is backed by a substantial sloping bank of earth and rubble, which increases the thickness of the dam at its base to 22 feet.

A distinctive feature is the outlet, which is built in the dam and was made to Mr. Marshall's own ideas, the general principle of which is shown in Fig. 2, and will be seen to be a clever contrivance to obviate much of the annoyance associated with the ordinary screened outlet. In the cross-section depicted, B is the gangplank above and across the outlet, and A is one of the inspection doors, of which there are two that together reach across the gap, hinged to permit them to be easily raised. C is the usual vertical screen whose special mission on earth in the ordinary type of outlet seems to be to get choked with undesirable flotsam. In the present style of outlet, however, it merely acts as a second line of defence and does not come into action under normal conditions. Instead, the waste water is drawn off from below the surface through horizontal perforated iron plates, D, passes through apertures left in the base of the

THE OUTLET OF THE TARN: GENERAL VIEW.

[*Facing page 72*

concrete partition wall, E, rises to its own level on the other side of the obstruction, and, tumbling over the outfall, slips away down the concrete and masonry chute to resume its original functions as a mountain stream. F is the concrete moulded to the requirement of the outlet.

No fish can play truant through the perforated plates, while the bulk of the floating scum and litter finding its way into the opening from the tarn rides harmlessly against the partition wall. In extraordinary circumstances, due to an extra push of water, or the partial or complete choking of the holes, if the perforated plates prove insufficient to permit the passage of the whole of the waste, the water rises until the vertical screen comes into operation. For anything beyond that—cloudbursts or special Lakeland rainy visitations—Mr. Marshall has provided an efficient spillway away from the dam so that all fear of the earthwork being washed away by the flood water is obviated.

In making an artificial tarn the constructor should never lose sight of the susceptibility of his pet creation to the attacks, if unprotected, of two insidious enemies—silt and starvation. The rapid mountain torrent, in its wild roystering moods of winter violence, carries along with it vast quantities of gravel, and would lay them with mischievous delight in accumulating mass on the floor of the tarn. Where such a stream is the feeder of a lake its disastrous propensities in this line may be to a great extent frustrated by the introduction of a small dam across the inlet, the pool formed thereby acting as a settling tank, which can, when necessary, be cleaned out. The danger of silt depends, of course, on the nature of the stream or streams flowing into the tarn. In some instances it is negligible.

The other enemy is the common foe to almost all artificial lakes holding a heavy stock of fish, though, in spite of the lessons repeatedly taught by the fishing records of great reservoirs, is too frequently forgotten until the lean years have set in and redemption is difficult of attainment. It is now an old tale, but its importance calls for brief repetition. For the first few seasons after a reservoir is made the trout in it thrive amazingly, rapidly assuming aldermanic proportions. Then the fishing falls off to an alarming degree, and from the status of a West End restaurant thronged with sleek and corpulent gastronomists, the water is transformed into a miserable workhouse haunted by cadaverous Olivers asking for more. The reason? It has been found that for a year or two the submerged land vegetation produces great quantities of fish food, then it dies away, and, unless provision has been made beforehand, there is not a sufficiency of aquatic, food-producing vegetation to take its place. Realising this, the preserver frantically attempts to make amends by dropping in stones to which are tied aquatic plants, in order that they may take root and spread. Had the correct course been adopted the bottom of the tarn would have been planted with aquatic plants before the water had been impounded, and they would have been spreading while the value of the submerged land plants was declining.

Undoubtedly lakewort is the most valuable plant for the bed of the tarn. It will grow in most waters, and is never a nuisance. Next to it comes the water lobelia. Of these two plants, one will often grow where the other will not, but by all means let lakewort be the first choice. Then about the head of the tarn there are pretty sure to be thin shallows where watercress and marsh marigold will flourish, and no attempt should be neglected to establish in the lake a generous store of the common water snail. When a tarn is carpeted with lakewort, its inlet half choked with water-cress and marsh marigold, and its depth crawling with water snails, it may safely, like the popular boarding-house keepers, put in its advertisements "A good table kept."

Three aquatic plants there are which should never be introduced into a lake or tarn—American weed, *Anacharis alsinastrum;* water milfoil, *Myriophyllum;* and pond weed, *Potamogeton.* The first two spread nearly as fast as scandal, and create about as much trouble. In the race for choking the water it is not far from a dead heat between them. The pond weed is a little slower, but about equally effective in closing the water to the angler if given time and opportunity. Once introduced, these vegetable pests are most difficult to keep in subjection, and can only be eradicated by the adoption of the most drastic measures.

Left to itself the artificial tarn, fed by a natural trout stream, would in time stock itself ; but that means its sporting possibilities would be held up for a few years, while even after that, if fished at all hard, restocking would have to be resorted to. Yew Tree Tarn exemplifies how quickly a newly-formed sheet of water can be made to yield good sporting results. Fortunately, the owner possesses an up-to-date private hatchery, with rearing ponds, capable of dealing with 100,000 ova per season, and after the completion of the tarn he turned into it about 35,000 two-year-old fish—Loch Levens, Rainbows, Irish trout and *Fontinalis* char—all reared in his own ponds. The third year he opened the tarn for fishing, when twenty-nine rods, fishing fly only, took 522 trout of 9 inches and upwards.

Supposing the prospective demand upon the tarn does not warrant heavy stocking, and that there are fair numbers of trout in the beck feeding the tarn—perhaps spread over two or three miles of stream—the question arises : What can be done to transfer these wild fish to the new elysium in numbers, instead of waiting for a portion of them to drop down into the tarn casually, dribbling in slowly ?

If there is a free run into the tarn, and if there is any reasonably rough approach to uniformity in the depth of the feeder stream for some distance above the inlet, the trout in that length may be swept down into the tarn by adopting the old Indian method of utilising the "brush-drag." The Indians used as the foundation of their drag a vine, but in the present instance three tough laths, long enough to reach across the stream at its widest part will be found handiest. The three laths should be fixed in position, about 2 inches from each other, by nailing to short cross pieces, say 2 feet apart. Having thus constructed the frame of your brush, you fill in branches of shrubs for the bristles—willow, hazel, or old heather if it is handy. Cut the twigs long enough to give them a good grip of the bed of the beck, and lace the butt ends of them between the laths, under the

"BRUSH-DRAG," SHOWING METHOD OF CONSTRUCTION.

lowest lath, over the middle one, and under the top one. The twigs may be further secured in position with wire twined round them and the laths as opportunity dictates. Proceed in this way until you have made a long flat besom, with the branches generously bestowed and interlacing one another. Also, at the ends let them fan out, so that, in operation they may bring their disturbing influence into effect at points where the banks overhang. If the banks are low the top lath may be longer than the others, a projection at each end serving as a handle.

Placed across the stream at the top of the length to be operated upon, it is worked down after the manner of a trammel, and, if closely constructed, it sweeps more thoroughly than a net, for it accommodates itself to holes and obstructions in the bed of the beck. When the trout have been swept into the lake a screen should be thrown across the inlet to keep them there until they have realised the superior food attractions of the tarn, after which they will settle contentedly until the next spawning season sends them up to the redds.

If, however, the trout cannot be "brush-dragged" to the lake through the presence of obstructions to that method, then the trout in the beck may be netted out in the manner employed when hunting the wild ova in the spawning season. A suitable spot having been selected, a purse or bag-net is set across the stream so that any fish passing down must enter its open mouth, the narrowing funnel being, of course, extended down-stream by the current. Then by splashing, wading or other means the fish are driven down into the net. Here, again, the "brush-drag" is capable of rendering good service and will prevent many an artful dodger from doubling upstream. On coming to a portion of the stream which is narrower than the length of the drag, the machine can be adapted to the lesser width by working it aslant.

When the net is emptied the fish may be placed in pails or other carriers for transport to the tarn, in which the carriers should be immersed and the fish floated out. Also it is worth while to remember that trout may be handled in autumn and winter with far less risk to the fish than in warm weather.

# CHAPTER XII

## THE MOORLAND RIVER

THE typical English river of considerable length and upland origin may, I think, be divided into three sections, merging into one another. In its lowest reaches, where it crawls lazily through lingering miles of lethargic deeps, it is the materialisation of the coarse fish angler's dream. In its middle lengths, where the exuberant influences of its early youth have not been entirely overcome by the lassitude of the lowlands, it is a mixed fishery, the leading characteristic of which is that when you are not carefully unhooking and returning out-of-season coarse fish while you are deliberately and of malice aforethought in pursuit of trout, you are tenderly restoring trout that will not be shooed away from your bait when you are, under legal restriction, only permitted to be interested in coarse fish. The remainder of the stream, in its course from its home in the heart of the highlands, alternating in rippling glides, merry rapids, occasional tree-fringed pools, and dancing fords—orthodox wet-fly water—is classed as a moorland trout river, and as such claims the first place of interest in the sporting affections of a large proportion of game fish anglers plying both fly and bait in their seasons.

The moorland river presents to the angler, in a modified or extended degree, many of the problems of the mountain streams of whose combined waters it is formed (besides introducing problems of its own), but it cannot always be dealt with in the same way. From the angling point of view it resembles somewhat the precocious adolescent offspring—its case calls for a certain amount of corrective treatment, but it is a bit too big to be spanked with a slipper.

Kind and gentle it may seem when, under summer's shrinking spell, it prattles over the pebbles from shimmering pool to pool; but suddenly 50 or 100 frenzied becks fling into it simultaneously the devastating drainage from a cloud burst on the hills, and a raging flood sweeps down the dale. I have seen it come tearing down, a foam-streaked bank of solid water, and I have scrambled out of the river as fast as Providence and a fairly athletic constitution would permit me. I have watched a moorland river rise a yard in fifteen minutes, and I have seen 7 feet of flood water racing through the arch of a stone bridge where the stream an hour before was little more than ankle deep. I have been fishing in dead low water when, without any warning other than the roar of the fast approaching tide, the flood has swept away the trestles and tackling of a gang of men putting in stake-and-wattle protections for the bank before they could make any effort to save them. And never do I see or think of one of these sudden floods but I call to mind the sorry plight of an angling companion of a quarter of a century ago, who was marooned by one on an island all night, and then only escaped by wading breast-deep. You cannot do anything with such trifles as crude boulder dams against such forces—you might as well emulate the efforts of Mrs. Partington to sweep back the Atlantic with a mop.

These rivers, left to the unaided exercise of their natural reproductive powers, may have been equal to the modest demand made upon them by our sporting grandfathers and the happy band of poachers who pursued their little game by stealth and—at rare intervals—blushed before the magistrates to find it fame. To-day the case is far different. Rail and road transitional facilities, the growing popularity of the sport, the increased amount of leisure of all classes, wider recognition of the fact that by combination preserved trout fishing is within the reach of the working man club angler—all these have combined to let loose upon our moorland rivers battalions of trout fishers where erstwhile they came as single spies. The question of supply and demand enters as closely into the administration of our sporting fisheries as it does into our commercial economy, and the fact that the demand on the trout fishing capacities of our moorland rivers has increased fifty-fold or more calls aloud for something to be done. Something has been done; something is being done; but something more can and must be done if the sport is to be maintained at, or raised to, its highest degree of efficiency.

A FRENZIED BECK.

[*Facing page* 82.

Complaints are rife. They are made confidentially and do not usually find their way into the " Reports from the Rivers " published in the Press. One man bemoans to you that whereas twenty years ago he could frequently get his limit of ten brace in half a day's fishing in May and June, he has only touched the limit three times during the last two years, and then he has had to work very hard for it. Another admits that there are plenty of fish in the length he fishes, but of late years they have sadly deteriorated in size, while yet another deplores the apparent fact that in certain waters the trout have forgotten how to rise. Corroborative evidence supports these plaintive protests. Logically, all such effects must have their causes, and such causes are not now so far to seek nor so difficult to remedy—at any rate in some measure—as they were a generation or two ago when pisciculturists were groping for the light through a mass of misdirected enthusiasm. We have learned much from the successes of the pioneers, but, I modestly uphold, we have learned more from their mistakes and failures, for it is often more valuable to be certain what not to do than to possess a merely moderate assurance of the correct thing required. Anyhow, while there is still much that is hidden in the solution of the problems of the moorland river in what I have heard described as the test-tube and microscope department, the way lies wide open to big improvements, and only the requisite energy is called for to effect such improvements.

These improvements may be classed under two heads, viz., those which depend upon the actions of the lessees or riparian owners, and those which are within the province of the legally appointed boards and councils. With regard to the latter, the owners and lessees can do little beyond worrying the powers that be into a better appreciation of their duties and the utter futility of Acts and By-laws that are not enforced. Theoretically our fishing streams are protected from pollution. Practically, especially in more or less remote districts, pollution continues to be practised openly and with impunity. I have myself seen the handy man of a Dale village pasting up authoritative posters threatening with all the pains and penalties of the law any miscreant who deposited rubbish in the stream. His very next job was to tip into the stream barrowloads of what he called " refuge " at a spot within 10 yards of where he had just put up his last warning.

Again, the tar-spraying evil has undoubtedly been responsible for incalculable mischief in the form of direct destruction of fish and still more in the shape of the wholesale wiping out of aquatic plant life and the trout food produced thereby. The county councils have been compelled to recognise the presence of the danger, and in some instances have given assurances of the issue of orders that at points approximate to fishing streams an innocuous substitute for tar should be used on the roads. But are these orders in all instances carried out ? Witnesses are not lacking to affirm that they are not, and the condition of many tributary streams, receiving road washings and discharging them into the main streams, points to spraying methods as an ever present peril which cannot be too closely watched or too carefully guarded against.

I do not wish to have it appear that I am blind to what improvements have been effected or attempted. I believe I am voicing the feelings of all anglers when I say that we acknowledge with gratitude every step taken by the authorities in the direction of restoring to our streams their pristine purity, and we realise the difficulties that stand in the way of effective administration. But also, as anglers, we are convinced on the evidence of our own senses and experiences that very much in the fight against pollution remains to be won. We earnestly ask the powers to put forth that extra energy required to complete the good work of safeguarding our fisheries against chronic and incidental pollution and the risk of devastation. More, we pay tribute to the boards, we enrich, to the extent of the rates levied on those fisheries, the coffers of the councils, and we claim it as our right that we should be afforded the protection for which we pay.

I stress this point because without reasonable purity of water the most energetic efforts of the riparian owners and lessees to improve the sporting value of the river by netting out vermin, stocking, increasing the food supply, or in any other way, are hopeless from the outset. Fervent as the prayer that came to "The Haunted Man" should be our own incessant supplication, "Lord, keep our rivers clean!"

# CHAPTER XIII

## THE UNDESIRABLE ALIEN

In the preservation of the moorland trout river, no less than in practical political economy, it is essential to complete success to kick out the undesirable alien. Even if he is not to a marked degree cannibalistic in his habits, as a food filcher from a limited larder he is a constant menace and a real hungermonger to the trout. You cannot keep the boarding-house table well supplied when the neighbourhood is infested with a devouring horde of marauding scalliwags and the pantry door is left unlocked.

It may be natural for the mind attuned to the sense of proportion to accept unthinkingly the belief, " the bigger the river, the bigger the fish," with the consequent deduction that, proceeding up-stream, as the water diminishes in volume the trout dwindle in size. This is true enough of some moorland rivers, but it can hardly be admitted as a reliable guide. Other factors than the size of the river have to be taken into account, and one of the most important is the presence of other varieties of fish in the water. Often enough the trout average best in the lengths above the point where they have to scramble for the means of existence with coarse fish or grayling, and a marked improvement in the condition of the trout will follow the elimination or drastic reduction of, say, the chub in a fishery haunted by that fish.

Fortunately the moorland river is not, as a rule, the scene of the ravages of the piratical pike. The streamy shallows and rapid runs are not to the liking of the bold buccaneer. Still, in odd instances, in the lower trout lengths, where long slumberous deeps intervene between the stretches of swift current, pike may have established themselves, and one of the keeper's perennial duties is to wage eternal warfare against them. On the principle that any stick is good enough to beat a dog with, any effective system of spiflicating the enemy without damaging the innocent is to be commended. Set trimmers for him in the deeps; shoot him as he basks in the summer sunshine; seek him in spring in the ditches and shallows where he has retired to spawn, and wire him there if you cannot jab a spear into him. You are fighting him all in, Lancashire style with clogs, rough and tumble, catch-as-catch-can, kick and gouge, and the Marquis of Queensberry's rules have for the nonce gone into the discard.

Instructions for trimmering have been too often given to necessitate recapitulation here, but the details of wiring are not so generally known, and as, in the hands of an expert, wiring is one of the most effective aids to the extermination, or decimation, of the enemy, it is worth a brief notice. The wire itself is a running noose of single or double brass wire selected as to strength with regard to the size of pike likely to be encountered. The size of the loop is adjusted likewise, *i.e.*, it must be a little greater than the girth of the prospective victim. The wire is securely attached to the fine end of a tapering bamboo of 12 to 15 feet in length, and the loop is held in position, in line with the bamboo, by being lightly clipped in a notch in the tip of the pole. Armed with this instrument of summary destruction, the operator moves cautiously along the bank until he spots his quarry, when he freezes into immobility at a point level with the pike, save that through his agency the wire-armed end of the bamboo creeps out and out as if it were the branch of a tree growing very fast, until it hovers over the surface of the water a little ahead of the fish. Proceeding as gingerly as if he were tickling a trout, the wirer now dips the noose into the water, and v-e-r-y gradually moves it downstream until it has passed over the head of the pike. If the fish be small enough to be lifted, the wire should engirdle it amidships, but in the case of one whose bulk precludes such a possibility, the desirable position is round the shoulders and behind the pectoral fins. In either case, once the wire has been manœuvred to the correct spot, a sharp backward pull releases the loop from the notch and the noose draws tight on the fish. Then the small pike comes flying through the air to the bank, or the big one is dragged pully-haully ashore with the noose tightly round his shoulders and " under his arm-pits."

The question has been raised as to whether, in

WIRING PIKE.

[*Facing page* 89.

wiring pike, the loop should be lowered into the water above the fish and slipped down over his head, or introduced below the fish and drawn up over his tail. Both ways are practised. On the southern chalk streams I believe the former method finds most favour, while elsewhere many wirers adopt the "tail-end" method. I vote for the "headlong" system because in such case if, through incorrect judgment of the exact position, your wire touches the fish it slips easily past obstructions and the fish does not seem to mind a slight tickle; but if you proceed from the tail end, you are stroking the brute the wrong way and the wire will not pass the fins. Also, if the fish is lying over a weed-bed he lies with his head upstream and the weeds incline downstream with the current. Therefore, if you chance to include both the tail of the pike and the tail of a weed in your wire, and attempt to pass it up over both, you stand a very fair chance of wiring your weed and losing your fish, whereas, operating from the other end of the pike, your wire would pass between fish and weed. As to scaring the pike by the head-first method, there is little real danger of that if care is taken. Occasionally the pike may give some sign of slight annoyance, but it is generally in the form of moving off a little and resuming his doze.

In a few cases, too, I have known these lower trout lengths invaded to a troublesome extent by barbel, and I recall weird nocturnal scenes on the river when the barbel were in the streams and on the fords, and organised raids were made upon them by "blazing the water." Save that no boats were used, these raids vividly recalled the scenes of "burning the water" for salmon depicted in "Guy Mannering" and Scrope's "Days and Nights of Salmon Fishing," the dancing, flickering lights of the torches and flares strangely lighting up some of the figures and trees, while throwing others into deeper silhouette or half-burying them in more sombre shadow. Armed with spears, the little band of waders moved upstream, each attended by his torch-bearer, peering into the water for the dark form of the barbel hugging the bed of the stream, while other acolytes gathered up the spoils as they were speared and thrown to bank, and deposited them in sacks. Though "blazing the barbel" resulted in wholesale and wholesome destruction of that form of vermin, it has, so far as my knowledge goes, fallen into desuetude. It seems to me that, with the advantage of modern illuminating facilities, in the way of electric torches and lamps, superseding the old torches and flares, the practice might with less trouble and equal effect be utilised in those trout fisheries where the barbel have not yet ceased from troubling.

By far the greatest coarse fish problem with which the moorland river preserver has to deal is the chub question. The chub are to be found at much higher points of the stream than other coarse fish of notable variety; they breed prolifically; they absorb an immense amount of food, putting the trout on famine rations, and if unchecked they would, practically, squeeze the trout out of existence. Nothing but drastic netting will keep them down, and, happily for the preserver, their natural habits render them easy victims to this form of depopulation at a time of the year when the operation can be performed under the most agreeable conditions. Compared with netting vermin out of a chalk stream in the cheerless chill of the late or early months, netting on a moorland river is a joyous summer picnic. For it is during the heat of July and August—and I know one man who is firm in the faith that the English year is made up exclusively of July, August, and winter—that the chub are gently cruising in great shoals in the shallow streams or holding big family gatherings on the gravelly flats. At such times I have, with the upstream worm, fished as for trout, thinned their ranks by many stones and been repaid with good sport for my trouble, but such methods are feeble beside what can be done with the net.

Incidentally, it has often been deplored in the angling Press that the coarse fish netted out of a trout preserve should be, for want of some go-between to make fitting arrangements, thrown away when they would be heartily welcomed as additions to coarse fish waters. That such a state of things should exist is a profound pity, for what is one man's pest is another man's *poisson*, as it were. Now that salmon fishery boards have taken under their wing the coarse-fish angler's interests, they might do much to avoid this waste, following the example of the Yorkshire Fishery Board, which has nets. These can be obtained on loan by clubs making application for them, as also can a large tank of about 2 yards square, which can be placed in the water for the reception of the netted fish so that they may be shipped to other waters where trout are not preserved.

So far I have said nothing of grayling, which, of course, abound in many of our moorland rivers, often almost throughout the whole length of them. Whether they should be cockered as cronies or cursed as criminals is for the riparian owners or the lessees to decide. My own experience of trout and grayling fisheries has led me to the conclusion that in certain streams the two varieties do fairly well together, the late autumn and winter sport provided by the grayling perhaps compensating for some slight inferiority in the trout in size and condition ; in other waters the grayling, if not kept down, rapidly outbulk and outnumber the trout, with the consequence that the latter, habitually relegated to the second, and precarious, sitting-down, suffer accordingly. In such cases, at any rate, the grayling should be netted, and they, too, facilitate their own destruction by haunting the gravelly shallows in spring and summer, when they may be taken with the trammel after the manner described in the next chapter.

## CHAPTER XIV

### NETTING OUT CHUB AND GRAYLING

IF the enthusiastic tyro would graduate in the arts of netting a trout stream, then let him seek out for his tutor a poaching veteran; with honeyed words get on the soft side of him, unloose his tongue with a corkscrew, and then, sitting humbly at his feet, drink in the words of wisdom that fall from his lips. There is no better way. Of course, you have first to catch your poacher.

An old angling friend of mine, intent in his younger days on starting a crusade against the chub, spent a night in a thatched cottage with three old poachers, one of them nearing his eightieth year. After the first bottle of whisky was finished another was sent for, and before the old sinners sped the parting guest they had given him their nets and promised him a practical demonstration. They were as good as their word, and by way of giving him a thorough grounding in the art, they allotted to him the end of the net that traversed the worst water. Sometimes he was wading, sometimes he was swimming; but he stuck to his end of the trammel, and what he does not now know about netting out chub is a negligible quantity. The first year over 1,000 chub were taken out of the fishery, and that was the commencement of a systematic campaign against the intruders which is pursued to this day. So thickly do the chub flock to the shallows in the height of summer that frequently upwards of 100 are accounted for at one draught of the trammel, while on one occasion in the fishery I have referred to the netters landed 350 at one scoop.

In the event of a competent poacher being unavailable, and in the absence of ocular evidence as to how it is done, the novice may garner sufficient knowledge from the written word to enable him to make a successful *début*. I know of no angling book which gives a description or working principles of the operation to be performed or the tackle required. Halford, in his "Making a Fishery" certainly has a capital description of the tackle for, and method of, netting out vermin in a chalk stream, but these were not intended to apply to work on a moorland river, where the conditions are very different. Here, having selected the piece of water to be netted—say a gravelly flat varying in depth from 1½ feet to 2½ feet, with, at the lower end of it, a shelving pebbly beach where the trammel may be drawn—a stop-net is set tightly across the river immediately below. The object of this net is simply to prevent the disturbed fish from escaping downstream. It may be a plain seine, well weighted at the foot to hold it to the bottom of the river, and corked to buoy the upper edge. This is the simplest form of stop-net; but a purse

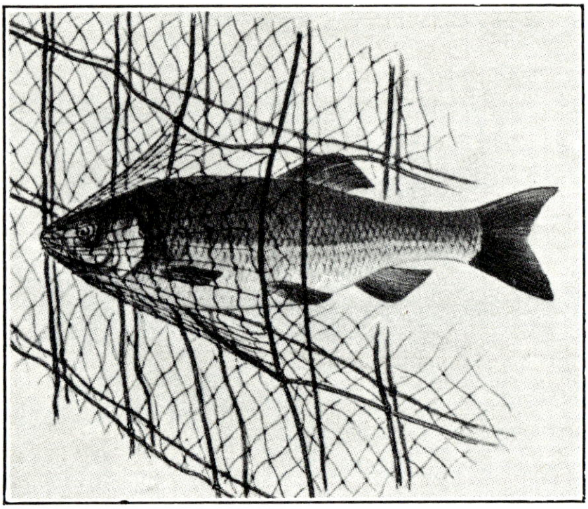

CHUB STRIKING THE TRAMMEL.

[*Facing page* 95.

or bag net may be employed, which is shaped like a funnel, the funnel being extended downstream by the current.

Now, in specifying the size of mesh in a net some confusion is apt to arise, to avoid which the measurements I give are from knot to knot, *i.e.*, when I refer to a $\frac{3}{4}$-inch mesh I mean a mesh that when open is a square $\frac{3}{4}$ inch by $\frac{3}{4}$ inch. This is a useful size of mesh for the stop-net. The other net to be used is a trammel, which is a rather more complicated affair, being three nets in one, sandwich fashion. The three nets are joined at their upper and lower edges. The two outer layers are known, according to locality, as the " walling " or " arming." They should be of strong hemp, 4-inch mesh, 1 yard deep, and should be long enough to reach across the river where it is to be netted. The meshes of the two outer walls should correspond with each other in position.

The inner net of the three-ply sandwich, called the sheeting, should be of fine hemp—what the poachers call " grey thread "—and its mesh should be $\frac{3}{4}$ inch. One of its essential features is the great amount of slack in it. Indeed, I have known the sheeting to be 4 yards wide where the walling is only 1 yard, and 3 yards is quite a usual allowance, and a good deal of extra length is also puckered in. Out of the water this slack hangs in accumulated folds at the bottom of the combination, but when in the river the buoyancy of the water effects a useful distribution of the slack between the two outer walls. The object of this arrangement is to permit the fish to pouch itself securely when it strikes the trammel. Alarmed by the general disturbance, a fish here and there blindly makes a dash at the net, with the instinctive intention of forcing his way through it as he would through a tangle of weeds, and falls a victim to misplaced confidence. For the moment he strikes the sheeting his further progress carries an enveloping amount of the slack with him through the large mesh of the outer walling, and he is helplessly imprisoned in the purse thus formed.

Corks of a size similar to that of a champagne cork may be bored and strung on a line of strong hemp, one to each alternate mesh of the outer walling, and secured to the upper edge of the trammel with fine hemp by a running stitch. The egg-shaped leads, say one to every third of the larger meshes, may be cast in lead in plaster of Paris moulds, should weigh about 2 ounces each and be secured to the bottom edge of the trammel.

Having fixed the stop-net in position, the trammel is put across the river at the head of the flat to be netted, and is worked steadily down-stream by a man at each end, while two or three men wade down behind the net to lift it over obstructions or render any incidental aid. These men should be well acquainted with the bed of the stream, especially if it is piled, for the same piles which are designed to tear the nets of the poacher show no favour towards those of the legitimate netter. Also assistants should beat and splash the water to drive the fish away from the banks—

in which vicinity there are possibilities of their dodging round the ends of the trammel. For the most part the scared fish move downstream in front of the net, but now and then the men behind the net see a sudden flash of white among the meshes where another struggling fish has found himself in the toils.

About 10 yards from the stop-net the end of the trammel on the beach selected for the drawing is brought to a halt; but the end at the other side of the river continues its progress downstream until the stop-net is reached, when it is brought across, immediately in front of the stop-net, until both ends are on the same bank and the trammel forms a horseshoe, encircling the fish it has shepherded down. Now, as the net is brought further towards the bank, and the decreasing area enclosed becomes, as the Irish satellite observed, " stiff wid " the bewildered crowd of victims of the raid, the scene becomes an animated one. Men splash about the congested area, literally chivvying the fish round in the endeavour to drive them into the net, to which they often show a marked aversion, while some of them are apparently so bemused they may be picked out of the water by hand. However, at length the trammel is dragged ashore; what trout have been incidentally captured are carefully restored to liberty, while the chub are dealt with as may be expedient.

Grayling, of course, are found among the victims, and go the way of the trout or the chub as the verdict may be. If, however, a special raid on the grayling is to be made, it is as well to net for them, in addition, in April or May, when they are spawning on the gravelly shallows.

Often—especially when the netting has been over uneven ground—the novice has stared with amazement at the relative numbers of the different species of fish brought to bank at the final draw of the trammel. Many a time I have known the net bring in nothing but chub, although the water netted has held a fair number of trout. The reason is that the summer chub swim high in the water, while the trout and grayling lie low on the gravel among the stones, and often when, say, the passage of the trammel over a biggish boulder momentarily opens a free passage you can see the trout or grayling streak through it, one after the other, like a flock of sheep through a gap in a hedge. The chub, in the higher stratum of water, fail to see such an opportunity to make their get-away, and are either pursed in the net or shepherded down stream. Consequently, while an occasional gap at the bottom of the trammel is not of vital importance when chub only is the quarry, the greatest care should be taken to sweep the bed of the river closely when the grayling is the object of the operation.

Finally, on the subject of netting, here is a word of salutary warning: Never commence your season's netting operations until you have carefully examined your nets and found them in sound condition. Otherwise things are liable to happen that, while relieving the drab monotony of an orderly procedure according to plan, are intensely irritating at the moment. I vividly recall an occasion upon which I was assisting a friend to net out the chub from his length of typical moorland river. The stop-net was set, each member of the little band was in his appointed place, light-hearted youth was venting the exuberance of effervescent spirits in lustily scaring the fish from the banks. We had got the trammel across the head of the flat; the keeper and his underling were operating the respective ends of the net; the owner and I were in the river following it down. We had got the drag nicely under weigh when my friend suddenly and excitedly yelled out: "Look out! Neck 'em! The net's rotten!"

And so it was—as far as the finer sheeting went. The first chub to strike it went clean through, and thereafter, keeping our eyes peeled, we were rushing from point to point, as fresh fish pursed, cracking their spines before their struggles burst the meshes. By the time we dragged the trammel ashore the sheeting was in tatters, and we only accounted for forty-seven chub, where had the net been sound we should have had twice as many. *Verbum sat sapienti.*

# CHAPTER XV

## STOCKING WITH FOOD

WHEN Dr. Johnson was shown a tempting menu as an inducement to him to attend a dinner, he replied gruffly, " Hang your bill of fare ! Show me your bill of guests ! " The order of precedence in the scale of importance on the moorland river is distinctly anti-Johnsonian. Make the bill of fare sumptuous enough, and you will have no difficulty about showing a bill of irreproachable guests. Having got rid of the undesirable alien, the remaining trout, with the entire contents of the larder to batten upon, might be content with a policy of *laisser faire*, but the preserver should not be satisfied until he has replaced the evicted vermin with additional trout and has insured an adequate food supply for all. How this may be done is our next consideration.

The pisciculturist's usual leading advice to put in aquatic plants is not applicable to the rapid moorland river any more than to the mountain stream. I have seen it attempted time after time, but I have never known it to succeed. The plants are ripped out by the first flood. The only really serviceable aquatic plant capable of withstanding the upland torrents is the water moss. This harbours a great amount of insect life, and is usually ready enough to establish itself on the boulders of a rocky river. Watercress, however, may often be grown in profusion in suitable places off the main stream. It frequently happens that a small stream, after tumbling headlong down the hillside, meanders sluggishly for a few hundreds of yards among the alluvial holms that fill the trough of the valley before emptying into the river. In this watercress flourishes profusely, and discharges a liberal and constant stream of food into the river. I know many such sites, some by natural growth, some by cultivation, which are wellnigh choked with the cress, and no better conditioned trout are to be found in the river than those in the parts adjacent to these shrimp-laden tributaries. To my way of thinking, the three most valuable flesh-forming trout foods that can be cultivated in a moorland river are fresh-water shrimps, common water snails (*Limnæa peregra*), and Stone-flies. After these the Caddis-flies render substantial aid.

Fifteen years ago, in an article in the *Field*, I strongly urged the claims of the water snail as one of the solvents of the food problem. Trout take it greedily and its reproductive powers are prodigious. I have opened fish caught in a river where *L. peregra* abound with their maws simply stuffed with snails. It is worth noting that this particular river, above the point where these snail-crammed fish were caught, has several deeps, and in the light that has emanated from the experiences and experiments of recent years, I am strongly inclined to the opinion that in a moorland river, subject to heavy floods, occasional deeps, natural or formed by dams, are essential to the cultivation of this mollusc. Dr. J. C. Mottram has devoted close study to the subject of snail farming for trout food, and devised simple but capital methods of promoting propagation while protecting the propagators. Writing in the *Field*, he says :—

"In order to promote an abundance of snails, protection should be given to the full-grown ones either laying eggs or about to lay them. This protection of egg-laying individuals is especially necessary because the fish appear to find the full-grown ones easily and the little ones can readily hide themselves among stones and weeds. Making heaps of brushwood in the water and putting the snails in these was tried, but it was found that the snails soon wandered away from the protection of the brushwood and, no doubt, came under the attention of the fish. A better plan was found to be to take a wooden box, such as a packing case, and bore holes in the side small enough to prevent the escape of the snails, then sink it in the river by placing large stones in the bottom, so as to leave the sides above the water level. The snails will lay their eggs on the sides of the box, and the young ones, on hatching out, will crawl through the holes and distribute themselves in the river."

Dr. Mottram's experiments were conducted on South-country chalk streams, not subject to floods, and in the event of his breeding-case being installed on a moorland river, the top of the case would require to be covered—say, with perforated zinc—and secure anchorage assured on a site sheltered from the fierceness of the current. The snails may also be cultivated in ponds or stews fed by a trickle of water, off the main stream, as suggested in dealing with the food problems of the mountain stream.

On this method of culture I am permitted to quote from a valuable contribution by Dr. Mottram to the *Salmon and Trout Mazagine* :—

" All that is required is a small stew fed by a trickle of water ; a large supply of water is a disadvantage because it promotes the deposition of mud. The sides and bottom of the stew should be thoroughly cleaned before the snails are put in. No water weeds are required, as the snails will feed upon the microscopic algæ, diatoms, etc., which will quickly grow upon the sides and bottom of the stew. The stew should be filled with flower pots laid mouth down and reaching to within a few inches of the surface of the water, these pots providing a large area for the growth of algæ and a surface upon which the egg masses can be deposited. The snails like to lay their eggs near the surface of the water, hence the necessity of having the pots reaching to near this level. Such a stew should be stocked with only one species of snail. Individuals of any size will do, but results will be hastened if full-grown individuals are used and if the cultivation be begun in the spring rather than the autumn. The only enemies to be feared are birds, especially ducks, either wild or tame ;

blackbirds and thrushes will also take snails along the margin of the water. When the stew is full of snails then the stocking of ponds or streams may be begun. This is most easily done by transferring the flower pots with the snails and egg masses clinging to them. The removed pots should be replaced by fresh ones, on which fresh eggs will be laid. In muddy waters it is best not to transfer the eggs, which are liable to be smothered and killed in the mud. By shaking the flower pots in a bucket of water the snails are washed off, leaving the egg masses intact and unhurt. . . . In stocking it is best to place the snails in shallow bays and side ditches, away from the trout. In lakes shallow bays may be netted off with wire netting to keep the trout away ; these will act as reserves for the snails."

Our knowledge of all the requisites for successful snail farming is not yet complete. It is apparent that the creatures do not like certain waters. Why, we are not quite sure ; but the fact that they are seldom found in quantities in the head waters of our moorland rivers may possibly be connected with their distaste for a peaty flavour in their tipple, acidity, or lack of lime. On the other hand, the fact that such waters offer little harbourage against floods may in some instances have something to do with the matter.

Persistent efforts on a fairly large scale to supplement the natural food supply of a moorland river have been in operation for some years on the Derwent Anglers' Club water at Forge Valley, near

Scarborough, and in order to present a reliable impression of the results, I got Mr. W. J. Clarke, the well-known Scarborough naturalist and keen angler, to give me a report. He says: "Since 1912 the Derwent Anglers' Club have stocked the river with various species of snails, fresh water shrimps and caddis grubs, in order to try to establish a better food supply for the fish. The experiment has not been an unqualified success up to the present. Snails do not thrive in the river. We have put in large numbers of *Limnæa peregra* and smaller numbers of *L. stagnalis*. The latter entirely disappeared; the former may be found now in the stream in small numbers, but they do not increase as they ought to do. We also introduced some of the small cockle-like mollusc, *Sphærium corneum*, and also some of the large fresh water mussels, *Anadonta anatina*, but we do not find any trace of either now, yet the fresh-water mussel is very abundant in the Scalby Cut, which takes its supply of water mainly from the Derwent. Many thousands of fresh water shrimps have been planted both in the main stream and in the little tributaries running into it, but they do not thrive and have not established themselves. Caddis grubs have done well and are now present in the stream in large numbers, forming the staple food of the fish, which consume them in great quantities, swallowing cases and all. Practically every fish stomach examined contains them. Last year we planted 30,000 *L. peregra* and 5,000 caddis; in 1924, 30,000 *peregra* and 4,000 caddis;

in 1923, 2,000 caddis and 1,000 snails. The Derwent, as you know, is subject to tearing floods in winter and after thunderstorms in summer. Consequently it does not grow a great deal of weed, and this may account for the disappearance of the creatures." It should be added that one of our leading piscicultural experts has reported the water quite suitable in chemical composition to the existence of the imported guests.

The evidence strengthens my general convictions. The failure of the fresh water shrimps to establish themselves is quite in accord with the fact that watercress, both in the river and in the tributary streams, is chiefly conspicuous by its absence. The signal success attendant upon the stocking with caddis grubs confirms the contention that these insects possess the ability to withstand the scouring floods of rapid rivers, while the fact that natural food abounds in the lower waters of the same fishery, which are characterised by deeps and slacks, points to the necessity of sheltered harbourage for some forms of fish food as well as for the fish themselves, and such harbourages can be found in the small tributaries, deeps and backwaters, whence the food there produced finds its way gradually into the main stream without much risk of the source of supply being swept away by the first flood. This applies especially to fresh-water shrimps and snails.

In the adjacent Scalby Beck, which takes its water largely from the Derwent, not only does the fresh-water mussel abound, but the common water snail is also present in limited numbers. The value of these advantages is shown by the fact that the Scalby Beck trout run to a heavier average in weight than their Derwent cousins, while the epicure might find further satisfaction in the discovery that a considerable proportion of them are pink fleshed, the result, no doubt, of their molluscular diet.

Here is another instance of the effect of a liberal snail diet upon the fish in a moorland river. On the same fell in the Pennines, within the proverbial stone's throw of each other, rise two moorland streams, the one taking an easterly the other a westerly course, but both passing through the same class of country. With the grayling of one river I was on terms of intimate acquaintance—they ran about three to the pound—but with those of the other river I was not even on nodding terms. Then a friend made a three days' angling raid on the other river, and astonished the natives, himself, and presumably the grayling, for he made a record catch, of which I saw one day's bag. I was astonished too. The fish ran from ½ lb. to 1 lb. in weight, and they carried the evidence of their well-being concealed about their persons. They were bulging, like a boy's marble bag, with water snails, mostly *Limnæa peregra*, and provided a striking object lesson on what can be done with a generous supply of good food when it is "put into a good skin." True, they were grayling, but what matters it whether we do the sum in haddocks or herrings, trout or grayling, so that we uncover the basic principle? Snail shells, big or little, did not seem to have troubled them any more than his first crab bothered the ingenuous young man from the country, who admitted that, while he'd managed to get it down all right, he'd a sort of suspicion that the crust had been baked a trifle too hard. I can entertain no doubt that in the instance of an ordinary trout being put on an exclusive diet of water snails, an all-wise Nature would promptly meet the case with the provision of a gillaroo gizzard.

I think it is beyond all question that wherever snails can be established in bulk the trout improve in size and quality. Therefore I urge strenuously that wherever deeps or mill-pools intervene between the streams of a moorland river attempts should be made to colonise them with water snails, the cultivation of which should be energetically pursued in every communicating pond and sheltered backwater which offers opportunity.

I use the word "bulk" in connection with snails advisedly. For the snail bulks well, and food in the bulk you must have in your fishery, ever keeping in mind that it has been computed, in the case of the rising generation, 5 lb. of solid food are required to produce 1 lb. of trout flesh.

Besides the fresh-water snail and the fresh-water mussel, molluscs which form valuable trout food where they can be established are *Planorbis corneus*, the *Pisidium* cockle-like bivalve, and *Ancylus fluviatilis*.

Reverting to the proposition that snails do not

WHERE MILL POOLS INTERVENE.

[*Facing page* 108.

like certain waters, the extent to which scientific research has advanced towards the solution of this problem may be gathered from what follows. Coincidently, while much of the matter forming this book was in the course of serial publication in the *Field*, Dr. W. Rushton, biologist to the Salmon and Trout Association, and one of our leading authorities on the phenomena of life in our trout streams, was conducting laboratory experiments closely bearing on the subject of this chapter, and any value the chapter may have had is greatly enhanced by the fact that I am fortunate enough to be enabled to quote from the valuable results of Dr. Rushton's experiments as communicated to the *Field* :—

" Most observers agree that the chief foods of trout are, as pointed out by Mr. Platts, molluscs, freshwater shrimps, Stone-fly larvæ, and caddis, with the various larvæ and adults of the *Ephemeridæ* and smaller fry, such as the water fleas, thrown in. It is mainly with molluscs, shrimps and caddis that I wish to deal, basing my remarks on laboratory and field experiences.

" For some months I have kept under observation in aquaria specimens of *Limnæa peregra, L. stagnalis*, freshwater shrimp and caddis larvæ. . . . From day-to-day observations it is quite evident that the food of the snails consists of these unicellular green algæ and that they are necessary for the well-being of snail life and for the increase of this form of life. It is well known that neutral or faintly alkaline waters are necessary for good algal growths, and most anglers are aware that such algal growths as flannel weeds are more numerous in chalk streams than moorland waters. If one examines the algæ or slime on stones in a peat stream, one finds that the most common forms, when any are present, are of the filamentous variety and belong to the blue-green rather than the green variety, whereas the algæ of filamentous forms adhering to the stones in a chalk or limestone water belong to the green types and here there are more unicellular forms than in a moorland stream. The algæ in such waters are also found adhering to the higher green plants.

"In the spring the three glass aquaria referred to were set going, with a view to see what does happen when the three types of trout food are living in captivity. Into No. 1 were introduced fresh water shrimps; into No. 2, *Limnæa peregra*; into No. 3, *Limnæa peregra,* and a gelatinous ring of caddis eggs from reservoir where fish are very abundant (the caddis being *Phryganea grandis*).

"In the middle of October, when I write, the few shrimps have multiplied to about 100 and seem to find sufficient food in the green algæ which develops and multiplies on the sides of the vessel and on the mud at the bottom. The *Limnæa peregra* in No. 2 have also multiplied many times. In No. 3, with the *Limnæa* and caddis larvæ, careful watch was kept throughout the summer, and it was seen that, as soon as the caddis hatched out from the eggs, they at once made cases from the leaves of the water weed after which they began to feed. Then for a time they fed on the *Limnæa* present until those were reduced to only a few. It was interesting to watch the caddis larvæ pulling the snails out of their shells and eating them. And when this source of supply was all used up then the caddis turned on each other until only two were left and still survive.

"On the introduction of more snails the caddis at once attack and eat them so long as they are within reach. If the snails creep up above the water plants along the sides of the vessel they escape capture, but should they come within reach of the caddis they are at once attacked. When no further animal food is available then the caddis turn to a vegetable diet and live on the algæ and soft parts of the water weeds.

"The observations seem to point to the conclusion that, when the three forms of trout food (caddis, shrimps, and snails) are introduced at the same time, snails and shrimps must have small algæ on which to live, whereas caddis will prey on snails if they get a chance and only go for a vegetable diet when forced. So long as snails are accessible they will attack them till they are all used up. If what has occurred in the observation tanks occurs when all three types are put into waters deficient in algal growths, the facts may to some extent account for the disappearance of shrimps and snails in waters where all three have been introduced.

"On looking up what references were available on the habits of caddis I found that Vallisneri had observed caddis very carefully and was the first to give an account of them. He says: 'Caddis feed upon the leaves of celery, ranunculus, and water-dock, as well as on other aquatic plants. Caddis worms are usually vegetable feeders, but not exclusively so, as they devour the larvæ of gnats, *Ephemeræ* and other aquatic animals. Some are altogether carnivorous, *e.g.*, *Hydropsychidæ*. When hungry, caddis worms devour each other. They do not attack the end of the case, which can be defended, but tear open the middle of the case.' "

Of course, Dr. Rushton is careful to refrain from any dogmatic pronouncement that the phenomena observed in the tanks with their restricted conditions are bound to be repeated, *pro rata,* in the river with its wider resources. Nevertheless, the lesson to be learned seems quite clear, viz., water free from acidity produces green unicellular algæ; green unicellular algæ produce snails; snails produce lusty trout. True, there is the intervention of the caddis worm to be considered, and in this respect it is comforting to know that while the caddis only breeds once a year, *L. peregra* continue to breed, generation after generation, throughout the greater part of the year. From the fact that *peregra* are to be found in abundance in certain waters which are also inhabited by caddis grubs, it would appear that, given favourable conditions—green algæ and suitable harbourage—they are capable of breeding faster than the caddis worms eat them.

There is another hint of considerable value to the moorland river preserver in Dr. Rushton's notes. To establish the fact that a stream is slightly acid is not always easy. Where acidity does occur in moorland streams it may be fluctuating in degree or it may be intermittent, consequent on weather conditions, such as heavy rains after periods of drought. Out of half a dozen tests of the same water, conducted at different times, three may give acid results, while in the other three instances the verdict may be "acidity nil." Why not, then, adopt the practical botanical test for the chemical analysis. Does your water produce green filamentous algæ? If so, carry on.

I am thankful to say that very many of our moorland rivers and upland streams do not react to the acidity test, and are only prevented from growing flannel weed in abundant masses by the mechanical forces of the spates to which they are subjected. In many streams flowing from compensation reservoirs, and, consequently, free to a great extent from violent incursions of flood water, the flannel weed thrives, while it is a common thing, during a long summer drought, for it so to choke the attenuated streams of our bigger moorland rivers that it becomes a positive nuisance to the angler, until a welcome spate tears its ragged streamers from their moorings and washes them away. It would, then, seem to be the safe policy to cultivate snails in such waters as produce flannel weed and can afford harbourage for the molluscs, strong in the faith that if the caddis worms do take toll of the snails, well, the trout feed on the caddis worms, and so, as in the plot of the novel, it all comes right in the end.

It is worth noting, however, that Dr. Rushton's *caddis* v. *peregra* experiment was concerned, as to the first party, with the larvæ of *Phryganea grandis* only, and I think it has been recognised for a long time by pisciculturists and entomologists that *P. grandis* is the most bloodthirsty pirate of all his tribe, though I never dreamed that his tale of victims included the fresh-water snail until Dr. Rushton opened my eyes. That his enormities extend to trout fry when the opportunity offers is, I believe, fairly well known. Over thirty years ago, as described in his "An Angler's Paradise," Mr. J. J. Armistead conducted a series of tank experiments in which were respectively associated alevins and caddis worms, alevins and sticklebacks, and alevins and large beetles, and afterwards a mixed assortment of vermin. Mr. Armistead's verdict on the evidence of the tanks was as follows: "The conclusion arrived at was that the large caddis worms are the most destructive of all. The large beetle larva (*Dytiscus*) is a good second. They each settled minnows and sticklebacks by gripping them from below. The caddis worms seem to be more active and more persistent in the chase. . . . The moral of this is that it is unwise to have any but very small specimens in fry ponds."

As *P. grandis* is only one of the many species of caddis inhabiting our trout streams, there is as yet no cause for suspecting the majority of the species of being guilty of active "Peregracide."

## CHAPTER XVI

### MINNOWS AND MILLS

I HAVE found that much divergence of opinion on the minnow question exists among river preservers. The smallness of the trout in one fishery is put down to the absence of minnows; the smallness of trout in another fishery is put down to the presence of minnows. In the one case the argument is that minnows provide a substantial meal for the trout; in the other that the minnows are themselves devouring, often in its immature stage, the food of the trout. The truth would seem to be that where the river boasts only a meagre larder the minnows may provide a fish course for the adult trout at the expense of the fry, for it is upon the minute forms of food that should be absorbed in the development of the fry that the minnows flourish. The logical conclusion is that where a sufficiency of minute food for both fry and minnows obtains, then the minnows are a decided advantage, and *vice versâ*.

The danger lies in the extraordinary capacity of the minnow to multiply itself once it finds itself in ideal surroundings, and on no account should it be introduced into lakes. The risk is too great, for instances are not wanting of tarns whose fishing has been ruined by such an introduction, the minnows becoming masters of the situation through sheer force of numbers. In the moorland river, with its fast rocky streams, the danger is not so great ; but even in some of these, where deeps with sandy shallows at their edges intervene, the creatures breed so fast as to become a nuisance and an appreciable detriment, the worst point of which is that the little pests are so difficult to get rid of. Still, merciless war should be waged on them where they flourish too abundantly. Where the water is reasonably accessible approaches should be made to the live-bait dealer, who will net out what quantities he requires for his tanks ; also the fine-meshed minnow net should be in regular use by the keepers or the club members, while a few wire gauze minnow traps, set in suitable places, will sometimes, if emptied and rebaited at intervals, account for tens of thousands during a day's operations. As a last resort, though the remedy is an unorthodox one, keep ducks—with discretion. This is on the principle of setting a thief to catch a thief.

As a general rule, it would be the height of folly to give a flock of ducks the key of the fishery ; but there are special cases, requiring drastic treatment, where the services of ducks may be temporarily employed with the assurance that the special benefits they effect will outweigh the general damage for which they may be responsible. The time for their employment is summer when, in the minnow-infested streams, the little fish collect in dense shoals in the shallow water. I have in mind one particular spot where the duck *v.* minnow warfare was waged with consummate skill and strategy, for ducks are capable of great generalship. At the foot of an old stone weir was a shallow basin in which the minnows collected in tens of thousands. So closely packed were they that they reminded me of the black clouds of "sile" when the infant fry of sprats and herrings are inshore on the Yorkshire coast. Above, on the pool formed by the weir, lazily paddled half a dozen ducks. Suddenly, without any apparent signal, they fell into line and swam rapidly to the weir at the point where it joined the bank on the near side. Down the chute they slid, unobserved by the enemy, deployed well in the rear of the congested masses, cutting off their retreat downstream, and then charged through the panic-stricken hordes. Gobbling and splashing, they went through that scattering host like Gideon's band, destroying and sparing not, then sedately withdrew from the water at the opposite side, trudged back up the bank to the pool above, and paddled lazily about while they gave digestion a chance.

Gradually the scattered minnows headed back to the favourite basin at the foot of the weir and resumed their close formation, and in half an hour the ducks made another raid, following exactly the same details as before. And so the game went on. The minnows never learned discretion from their bitter experience. After each successive foray the remnants of the scattered tribes stupidly returned to the scene of disaster, their depleted ranks were filled by recruits from the waters below, and the routine of destruction went on almost with the regularity of clockwork.

Minnows suggest sticklebacks, but—fortunately, I think—these pugnacious little tartars do not love our rapid streams, preferring the sluggish shallows or the weedy ditch. There are, however, trout fisheries in which the truculent little tykes would thrive, given the chance, and it is possible that the question of their food value may arise. I know of only one trout preserve in which the sticklebacks are regarded as welcome guests—Blagdon Reservoir. Of their *locus standi* in the alimental economy of that famous fishery the manager, Mr. Donald Carr, wrote in *The Salmon and Trout Magazine*: "We regard sticklebacks also as an asset for big fish, and of these we had an abundance prior to 1921 when the bulk of our supply was pumped away. To remedy this we have restocked on several occasions, and it is gratifying to note there is now an appreciable increase. Sticklebacks frequent weed beds, shallows and ditches, and induce large fish to come inshore; and it is on these occasions the bank angler takes his toll." But Blagdon Lake is a thing apart; it possesses peculiar features, and what may be good medicine for its big trout might very easily prove the reverse for the smaller brethren of the average fishery. It is very well known that the stickleback is a remorseless foe to fry, besides absorbing their food; he is capable of rapidly multiplying himself into vast shoals, and the only safe course of the preserver of ordinary trout water is to keep him out of the fishery.

More especially with regard to the supply of larval and fly food I have heard the question asked on sundry occasions: Why do the chalk streams produce bigger trout than the moorland rivers? Beyond the fact that the crop grown on a mud and loam bottom must necessarily differ in quantity and variety from that produced on rock and gravel river beds, the chief reason is that, with fluctuations in volume, the chalk stream varies slightly in depth only, while the moorland river varies mostly in width. The chalk stream, flowing gently on with the almost imperceptible movement of a trades union labourer on day work, has its bed covered the whole of the time, that is, it is a food factory working double shifts, night and day. The changeable moorland river is continually expanding and contracting; it is in and out like a trombone; in its shallow reaches more than half its bed is dry for long spells, which effectually throws the spanner into the machinery of that portion of the factory and wrecks it for productive purposes.

I have already referred to this waste of what might be good food-producing ground in the case of the mountain stream and the importance of its reclamation. In the instance of the moorland river the situation is often much worse, the intermittent exposure of great tracts of river bed being aggravated by the operations of the mills which harness the current for motive power.

This is what happens in hundreds of cases in valuable English trout fisheries. The river is dammed with a high mill weir that backs up the water for, maybe, half a mile upstream. When the water in the river is below its normal volume the mill is taking it from the accumulated reserve behind the dam faster than the stream can replenish it. So long as the mill is running and the tail-race discharging into the river a little below the weir there is a fair volume of water through the fishery below the mill. But the mill stops for dinner, and the river stops, too! What water is coming into the dam from above is being impounded there for use during the afternoon shift. Below the dam, as the water drains off, leaving bare and baking in the sun three-fourths or more of the gravel bed, the river shrinks to a feeble trickle a few inches in depth and generally near one bank. Then, in due course, the mill resumes operations, and the river comes out again, maintaining a fair flow until, either because it is time to knock off or the reserve of water is exhausted, the mill shuts down, and for hours the greater part of the bed of the river below the dam is again bare, until the long, long pool above the dam has filled up and water comes over the weir.

These large areas of almost flat gravel, whether caused naturally or artificially, might well be the habitat of the lexicological schoolboy's amphibian —" an animal which cannot live on land and dies

POOL FORMED BY STEPPING STONES.

[*Facing page* 120

in water"—for they do not remain uncovered long enough to grow a terrestrial crop, nor submerged long enough to produce an aquatic one. All that is wanted to bring them into fruitful service is a foot-high barrage here and there; but how this is to be done cheaply and effectively is a question that has long worried me. At odd places local authorities or riparian owners have, unwittingly, effected considerable improvements by putting across the stream rows of substantial stepping stones. At first thought it would appear that such semi-obstructions would fall far short of producing such a result as the angler might desire; but in actual practice they are often the means of transforming that portion of the stream immediately above them into a wide, shallow pool, as shown in the illustration. Still, one cannot undertake the construction of a series of rows of substantial stepping stones which, in many instances, are but half-remedies, and it was not until I was preparing the matter for this book that the scales fell from my eyes and I recognised the solution to the trouble in the form of Mr. John Henderson's wire weir.

# CHAPTER XVII

## WIRE WEIRS AND FLY CULTURE

MR. JOHN HENDERSON'S system of wire dams, referred to in a note in the *Field*, appealed to me as the very thing wanted to effect the transformation of, in the aggregate, many hundreds of acres of almost barren river bed into food-producing areas.

Mr. Henderson has thoroughly tested out the practical value of his invention on the Dove in Derbyshire, which, in the technical sense, is a moorland river. The idea of using ordinary wire netting of, say, 1-inch mesh for damming a river may, at the first blush, strike the ribald as too ludicrously akin to carrying water in a sieve to be regarded seriously, but in actual practice it is found to be both economic and thoroughly effective, while the durability of such a barrage is proved by the fact that, constructed of ordinary rabbit netting, Mr. Henderson's first dam withstood the floods of the Dove for ten years before calling for repair. Even then the damage was not extensive. Metallurgists are, I believe, in hot pursuit of the extension of the "rustless steel" principle to iron for general purposes, and, once we have at our disposal wire netting that will successfully defy oxidisation, the life of the wire dam should be very materially prolonged.

THE WIRE WEIR, SHOWING METHOD OF CONSTRUCTION.

[Facing page 123

Both the general effect and the method of producing that effect are shown at a glance in the illustration to this chapter. A line of stakes is driven into the bed of the river at intervals judiciously determined with regard to the force and volume of the stream, and the netting is drawn tightly across on the upstream side of the stakes and fastened to them with staples. It is important that the straining posts at the respective ends of the weir should be firmly rooted, and these posts should be strengthened by anchoring posts, driven deeply into the banks at an angle, so that they incline away from the river and thus offer greater resistance.

The wire netting is bent at a right angle, lengthways, one portion being upright against the line of stakes and forming the dam, while the remaining width lies upon the bed of the river upstream and is buried in gravel. Taking advantage of Mr. Henderson's experience, we learn that in the case of netting of 5 feet in width the best results are likely to accrue if from 1 foot to 2 feet of the netting is fixed upright against the stakes, and the remaining 4 feet or 3 feet are buried horizontally. This suits our requirements admirably, for while we may make the weir 1 foot high across the higher portions of the intermittently exposed river bed, there is almost always a dip towards one bank where the upright portion of the netting will have to be deeper to ensure a level upper edge. Once placed in position, the netting is quickly choked by the flotsam brought down by the stream, the horizontal portion is further embedded, and the whole position is consolidated by a deposit of silt. Any danger of a local wash-out at the base of the dam, caused by the fall of the water over the weir, is obviated by the reserve of gravel-covered horizontal netting, which accommodates itself to any incipient hollow thus formed.

When you have secured the permanent submersion of these hitherto waste places—whether caused by natural conditions or industrial exigencies, or both—you will, in my humble estimation, have achieved the greatest practical step towards the solution of the local insect food problem. For, just as Nature will quickly cover a bare patch of earth with a luxuriant self-sown growth of vegetation such as clothes the adjacent soil, so she will quickly plant these reclaimed areas with a more or less luxuriant insect crop of the various species already inhabiting the river. Should it, however, be deemed advisable to supplement Nature's efforts, the greatest chances of success lie in confining one's attempts to those same, or closely allied, species. Of course, it is obviously useless to turn confirmed insectorial mudlarks on to areas where they could do nothing but stub their entomological toes on stony gravel.

Perhaps you have noticed on a length of river with which you are familiar that the hatches of March Brown have become sparser and sparser until that once common frequenter of the fishery has almost, if not entirely, disappeared from it; this gradual disappearance of the March Brown synchronising with the increase in numbers of the Green Drake in inverse ratio. Is that the signal for a vigorous attempt to bolster up the March Brown hatch by wholesale importation of fly ova of that species? No. It is a certain sign that the conditions of the river are altering; that, for some reason, it is no longer favourable to the life habits of insects of the March Brown class, but it is now offering the sort of home comforts particularly called for by the Green Drake class. Obviously it is the wisest policy to sow the kind of crop for which Nature has proved the ground to be most suitable.

Usually the transplantation of the *Ephemeridæ* is far from encouraging. In isolated instances the planting of the Green Drake has been claimed as a success on certain South-country rivers, and old John Fosbroke was credited with its introduction to the Manifold, but such experiments usually end as did that of one trout preserver who wrote me, "We have tried planting Mayfly eggs. We were informed on apparently sound authority that the fly hatched out in three years. We find this estimate errs a trifle on the optimistic side. We put in the eggs twenty years ago—and they haven't hatched out yet."

In another case about twenty bags of weed and mud from a noted Green Drake stream were planted in the waters of a club in the hope of establishing the fly in new quarters. The following season a limited hatch of Green Drake appeared. The next season the fly was up in such increased numbers that success seemed assured. Then followed some bad floods that scoured out much of the weed—and presumably much of the gritty mud required by the burrowing larvæ—and since then the hatch has dwindled and dwindled until at the present time it consists of a few odd stragglers.

The safest cards to play on the gravel-bedded moorland rivers are Stone-flies and Caddis-flies. The eggs of the former are easily procured during the hatching season on the banks of a stream where the fly abounds, where, on the dry stones or on the fences, the gravid females are to be caught in hundreds, heavily laden with big clusters of brown eggs which may be transported in water. Caddis grubs may readily be collected from streams in which they are abundant, or they may, as in the Derwent case, be procured from the commercial fishery. Of course, when the female Caddis-flies can be captured ripe for the depositing of their eggs they may easily be induced to surrender their fruitful burdens by dipping in water, and in the water the eggs may be transferred to the desired spot. A busy hatch of Grannom is a capital opportunity for this method of transplanting, the big bunches of blue-green eggs attached to the females being readily seen.

With regard to the cultivation of the Dun family, it seems to me that the greatest likelihood of success lies in, so far as possible, protecting the eggs of the flies indigenous to the river from the enemies that lie in wait for them on and about the river bed. Some time ago I had a nebulous idea that something might be done in this direction with floating rafts attached to swing hurdles, but that idea never got out of the clouds. In another quarter a couple of earnest caterers have been considering the notion of mooring branches of trees to stakes, which would probably have some beneficial effect in the direction aimed at—and would certainly, if enough branches were moored in fly-fishing water, prove a source of great emolument to a wide community of appreciative tackle dealers.

Meanwhile, "Middle Test," of the *Field,* has devised a simple, but eminently efficacious, method which he described in that journal on September 10th, 1925, giving full particulars with illustrations. Briefly, his plan comprises the provision of board rafts, moored to stakes or bridges, on which the Duns lay their eggs, and to which their natural enemies, including the caddis grubs, cannot gain access. On the underside of these rafts the flies deposit their eggs in thousands, and there they hatch out, the resultant larvæ, now in a stage better fitted to fighting the battle of life, making their way to their natural haunts on, or in, the bed of the river.

"Middle Test's" experiments have met with undoubted success on a South-country chalk stream. I am not very optimistic about the success of the raft incubators if they were placed in such a position in a moorland river that they were exposed to the savage violence of the floods, but I am confident that much benefit would result from their use if they were anchored in the more sheltered bays and eddies of the deeper reaches. Anyhow, the cost of the materials is so infinitesimal that there is everything to gain and nothing to lose by the attempt to adapt the method to our rapid rivers.

The Alder-fly certainly claims attention in connection with the present subject by reason of the facts that it is of all aquatic insects probably the easiest to transplant, and that it possesses a remarkable facility for making itself at home in almost any water in which it finds itself. Although in its mature, or winged, state it only comes within reach of the fish by accident, as when, for example, it is blown upon the water, it has a subaqueous larval existence, at two stages of which it is exposed to the gastronomic attentions of the trout and is much appreciated by them, hence its value in the fishery. The female Alder, a most orderly creature, deposits her eggs with mathematical precision in brown patches, each egg touching its neighbour, on leaves, rushes, grasses and other convenient flat surfaces near the edge of the river, and, preferably, overhanging the water. Very little search in the spring will reveal these ova deposits, so compact and neatly arranged in their brown patches that they cannot be mistaken. In a few days the larvæ are hatched, when they drop into the water, dive to the bottom and burrow in the mud. There they remain until ready to pupate, when they crawl ashore and bury themselves in the turf, emerging in due course in the winged state. It is when they are dropping in gentle showers into the water and later when they are making their way ashore that the trout feed on them, though, as the angler well knows, the fish will take with gusto the occasional mature insect that falls upon the water.

The clusters of eggs hatch in air, requiring no moisture, and may be sent through the post if due precautions are taken against crushing. The rushes or grasses on which they are found should be cut as long as convenient, then transported (with the eggs attached to them) to the fishery where they are required. Arrived there, all that is required is to fix them in their new quarters by any handy means, such as tying them to bushes, so that they overhang the water, and Nature will do the rest.

# CHAPTER XVIII

## HATCHING, REARING AND STOCKING

When the scientific trout fisher is not racking his inventive powers to devise new lures to render the taking of trout more easy, he is cudgelling his brains to frame new rules and regulations to make the catching of them more difficult. Somewhat similarly, what time the trout preserver's interest is not keenly focussed on getting the trout out of his fishery, it is acutely centred on getting them into it.

In the case of the average trout river, especially club water fished assiduously by a generous number of rods, the natural reproductive capabilities of the fishery are not equal to the task of supplying the demands made upon them, and recourse must be had to artificial stocking. You may purchase your fish from the commercial pisciculturist or you may breed them yourself; but in either case it is essential to the success of your scheme that you should, by means already indicated, have assured the newcomers a hospitable welcome, never forgetting that it is extremely unlucky to seat your guests thirteen at table where there is only enough food for half a dozen. In the case of turning out stock fish in the river itself, you will get the best results from two-year-olds, and it should be remembered that, whether procured from a commercial fish farm or from your own nurseries, these fish have been pond-reared and are not fitted to change their life at one fell swoop from the pampered seclusion of the sheltered rearing pond to the bustling battle of life of the rapid stream. You might just as well snatch the ducal babe from his plush-lined sanctuary and set him down to scrap for the daily crust in a busy thoroughfare among a horde of accomplished guttersnipes who are alive to every move of the game and have graduated as masters of artful dodgery. Turned out in the slack waters of the pools early in the close season, they will gradually accustom themselves to their new surroundings, and as they acquire the necessary strength to pull their weight against the current they will find their way into the streams and distribute themselves over the fishery.

Yearlings have little chance in a strange river. Brief life is there their portion. For the most part they are speedily taken in and done for by the older and piratical generations. If you must stock with yearlings, then let the fish be released in the small tributary streams, a good distance from their confluence with the river, and, where possible, interpose a screen between their nursery ground and the lower waters. If, say, through an abnormally heavy hatch in the boxes and consequent lack of accommodation in the rearing ponds for the entire crop, it is necessary to turn out fry, let it be done still higher up the tributary rills, where far fewer dangers lurk than in the river.

A wide world of uncertainty exists between the hatching of ova and the stocking of a river with sizable fish, especially when the operations are in the hands of amateurs. Millions of fry have been hatched out and turned straight into the river from the boxes, with the probable consequence that not one in a million survived that infantile stage. One enthusiastic, but misguided, amateur wrote me some time ago: "I put 2,000 fry into a nice shallow part of the river, where I could see and watch them. I stood as still as possible, and in a very little while bullheads and loaches began to peep out from under the stones by scores and take stock of the new arrivals, whom they speedily commenced to devour. They carried on the game until my heart was sore, and I came away fully persuaded that in twenty-four hours not one of the fry would be left alive."

During the earlier years of my acquaintance with club and private pisciculture it was the common practice to transfer the fry direct from the hatching boxes to the river. Then crept into vogue a change in the procedure. Instead of the fry being turned into the river to be promptly devoured, they were liberated in so-called rearing ponds to be starved to death by slow stages. Under this fasting system it was no unusual thing for a hatch of close on 100,000 ova to yield no more than 4,000 or 5,000 hunger-stricken yearlings, while if the census revealed the presence of 8,000 or 10,000 in the yearling ponds it was a matter for hearty congratulation. What better could be expected from a meagre bread-and-water diet in the administration of which the pisciculturist contented himself with providing the water and permitting the ravenous youngsters to skirmish round for the bread themselves? To-day there is no reason why the well-appointed club hatchery should not, out of 100,000 ova laid down, hatch out 95,000 and rear upwards of 50,000 fry to the yearling stage, after which there should be little loss in carrying them forward to the two-year-old mark.

To serve a well-thrashed, fairly well victualled fishery comprising from three to six miles of trout river of fair volume, a hatchery equipped for the treatment of from 50,000 to 100,000 ova will be found fully adequate if run in conjunction with a set of efficient rearing ponds. The furnishing of the hatchery, which need be no more than an unpretentious wooden building, is a subject which is fully dealt with in works on general trout culture, and I refer any one contemplating the starting of a breeding establishment to consult, if possible, Maitland's "History of Howietoun," Armistead's "An Angler's Paradise," "Fish Farming for Pleasure and Profit," by "Practical," the trout volume of the "Fur, Feather and Fin Series," and the article on Pisciculture in "The Encyclopædia of Sport," or consult a firm of pisciculturists who supply the necessary appliances. I would merely add that, of the various systems of laying down the ova in the boxes, the glass grille arrangement is the safest.

When visiting the new trout tarn near Coniston, I was shown Mr. Marshall's private hatchery near the head of Coniston Lake and immediately fell in love with it. It is exactly what a club hatchery should be. The hatchery itself, a wooden building, is situated in a fairy dell and has a capacity of 100,000. Immediately below it is the first of a series of rearing ponds, half hidden in a jungle of azalea, berberis darwinii, iris and other shrubs and plants, all contributing their insectorial additions to the menu. But that is not all. Above the hatchery are four food ponds, masses of watercress and other food-producing aquatic plants, from which the water is supplied to the rearing ponds. Thus the stream flowing into the nurseries is charged with natural food, and from their earliest infancy the fry, only partly dependent on the periodic appearance of the cloud of atomised liver and fish meal or, later, the bucket and spoon, are trained to look for the natural food in the form in which it will come to them when they are turned out into the world to forage for themselves.

It is not likely that the pond accommodation at the disposal of a club will be sufficiently extensive to permit the whole output of a 100,000 ova hatchery to be retained to the two-year-old stage before liberation. What proportion should be retained can only be determined by the circumstances surrounding each individual hatchery. The first year the whole of the ponds should be

A MODEL PRIVATE FISH FARM.

[*Facing page* 134.

engaged in rearing fry to the yearling stage. At the end of that period the ponds to be reserved as two-year-old nurseries are determined upon, and the yearlings in them are left undisturbed, while the remaining ponds are cleared of yearlings (which are distributed in the tributary streams) in readiness to receive the next complement of fry. After that the annual programme falls into something approaching a regular routine: Empty the two-year-old ponds and restock them with yearlings; empty the yearling ponds and restock them with fry, the surplus yearlings and fry being liberated in the small feeders of the fishery under the most advantageous conditions (for particulars see the text-books) which circumstances permit.

Ova for laying down on the grilles may be purchased from the commercial hatcheries, or the native spawn may be utilised, and I hold that only the man who has hunted the wild ova on a cold winter's day, suffering a thousand discomforts, and has stripped the fish with numbed hands in the icy rain, can take that whole-souled interest in the fascinating pursuit which the subject merits. The installation of a fish trap, however, in a favourite breeding stream obviates much of the physical misery attached to these wintry pleasures. An efficient form of such trap is detailed by Mr. Wilson H. Armistead in "Trout Waters," from which, by permission, I quote the following description:—

"I have always used a large wooden tank, built into the stream, taking care that the bottom is so laid that the current cannot in flood time work its way underneath. . . . The tank is composed of a wooden floor and two wooden sides, with screens at each end through which the stream flows. The screen at the upstream end may be a fixture ; but it is better to have it so arranged that it may be removed when the spawning season is over. The screen at the lower end should be made in halves and hung on hinges so that it may open inwards. When the trap is set, the two lower screens are slightly opened upstream, so that there is an opening of about a foot between them. To this opening and inside the trap is fixed a net funnel extending from top to bottom, and about 2 feet into the trap, reducing the aperture to about 3 inches. Through this the trout will find their way. . . . There are several ways of keeping the net funnel extended ; but I found the most satisfactory was to have an iron rod hinged to the screen reaching from top to bottom, with one or two arms projecting horizontally upstream, to which the net was laced. . . . The screens are best made of wooden bars well supported by cross horizontal pieces behind. For all ordinary conditions these bars may be cut $\frac{5}{8}$ inch by 1 inch and fixed with the narrow edge upstream. . . . It will seldom be necessary to place them closer together than an inch. Screens of this kind do not block readily, and if any rubbish should collect they are easily cleaned with a rake. . . . It takes two people to overhaul the fish in a trap of this kind, and the way it is done is as follows : Two nets must be provided, one on a square wooden frame which just fits inside the trap, and one landing-net on an iron frame, square or oblong. When the lids have been removed the hinged screens at the lower end of the trap are closed, and the square net on the wooden frame, which is known as the stop net, is lowered perpendicularly until its lower edge is resting on the wooden bottom, care being taken to keep the sides of the frame close to the sides of the trap. The stop net is then gently worked up-stream, without letting any fish past, till it is within 4 feet of the upper screen. Then, by means of wedges, it is held firmly in position. All the fish are now in the narrow partition between the stop net and the screen, and may be lifted out by means of the square landing-net."

If it is considered desirable to introduce new blood into the fishery, no method is handier or more economical than that of procuring the new blood in the form of ova from the alien hatchery and laying them down on the grilles.

Thirty years ago a prevalent popular notion ascribed the smallness of moorland trout to the long-continued breeding in-and-in of the fish, which, it was stoutly held, impoverished the blood and resulted in a community of stunted degenerates. Consequently the cry was, "Cross with other strains! Introduce fresh blood! Mongrelise! What livelier kicker can you have than a husky mule? What gamer fighter than a bull terrier? What more stimulating refresher than whisky and soda? What more mellifluous purifier than brimstone and treacle ? All tried and trusty mixtures. Mix your trout strains ! Ginger them up with the good red corpuscles of an alien race ! " Officials of one club which ran a hatchery met officials of another club which ran a hatchery, and exchanged ova with all the enthusiasm of philatelist fanatics swapping stamps. The golden age dawned on the professional pisciculturist. There were not lacking those who possessed and professed a deeply rooted faith in the theory that there is no water which cannot be improved by the addition of a little Scotch. The Scottish hatcheries got busy, and from that day, with characteristic national enterprise, great numbers of Scottish trout have annually come over the border to pick up a living in England.

As one who has fished much in trout waters where cross-breeding has been carried on on the higher scale, I am honestly of opinion that certain benefits have accrued from this intermixture of blood ; but these benefits fell short of the chief object aimed at. The trout remained small, as they must always remain small if they lack a sufficiency of food to plump them out.

Even if a hatchery and rearing ponds are out of the question, and your fishery has natural spawning tributaries running into it, much can be done to increase the yield of those tributaries by protecting and improving the natural redds. Every winter millions of ova are lost by the floods burying the redds under impenetrable piles of silt and by washing out the ova, as well as through the overcrowding of the redds, in which one pair of spawners wreck the propagatory work of their predecessors. There are often places on the stream where the spawning beds can be raked over previous to the run of the fish and a by-pass cut to carry harmlessly away the storm water. Occasionally, too, there are places where the stream splits, to reunite 100 yards further down. This offers splendid opportunity for the establishment of a maternity home and nursery. In either case, when a sufficient number of fish have spawned on the redds, stop further access to them by throwing a screen across the stream at some point lower down, and, if practicable, cover over the beds with wire netting. If it can be arranged so that the fry, after hatching, drop down into a watercress bed you are pretty sure of a good crop.

Or you may try some rudimentary form of artificial hatching with the use of hatching boxes placed in the stream. An ancient tin trunk will serve, and thus end its days in honourable utility. Punch the sides, bottom and ends full of holes like a colander, and punch from inside, so that the jagged edges are outward. Then partly fill it with clean gravel about the size of a pea. Next sow the impregnated ova on the gravel and cover with an inch of more gravel—carefully. Anchor the loaded box securely in the current, so that the stream enters the holes at one end and percolates freely through the gravel. Thus secured to a great extent from accident, the ova will hatch out, and the alevins, gradually working their way out of the gravel, will escape through the holes to take their luck in the stream.

# CHAPTER XIX

## ONE OF THE NEW DIFFICULTIES

For the purpose of providing the fish with places in which to find sanctuary from the violence of the roaring flood torrents, as well as of protecting the banks from erosion, the stake-and-wattle groin system, mentioned in reference to mountain streams, produces valuable results on the moorland river. Needless to say, such apparently frail structures cannot be extended to divert the main stream of any river of volume—the first flood would sweep them away—but short groins built slantwise into the edge of the rapid streams are capable of effecting astonishing changes. Let us glance at an example of what has actually occurred within my own knowledge.

The river at the point operated upon is of fair size, perhaps passing along 50,000,000 gallons of water per day under normal conditions. Out from the bank into one of the roughest streams were built some twenty years ago five stout stake-and-wattle groins, which are still standing. The stream was from 12 inches to 18 inches in depth and its bed was, typically, of clean-swept coarse gravel with submerged boulders embedded, shelving to the sides. What has happened will be apparent on glancing at the sketch plan. On the upstream side of the highest groin the gravel has silted up almost to the top of the stakes, while at the end

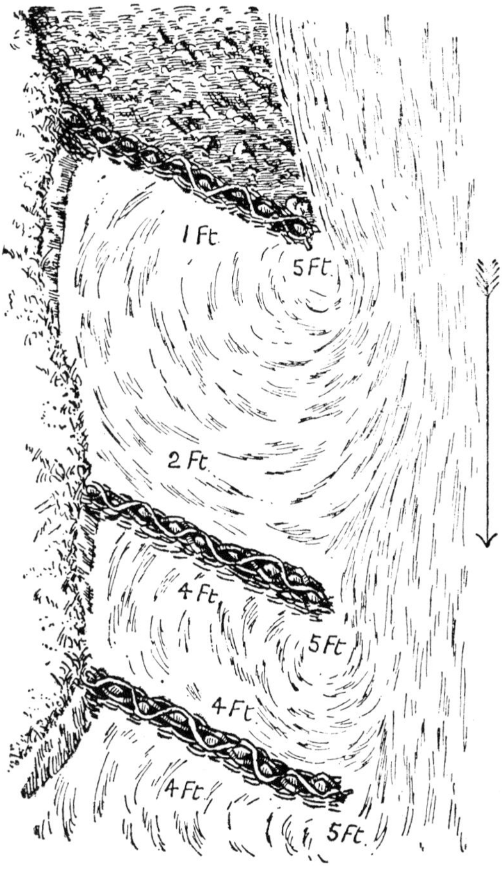


THREE OF THE GROINS.

of the groin on the downstream side has been scooped out a basin of 5 feet in depth. Between this basin and the bank is slack water varying in depth from 1 foot behind the first groin to 2 feet in front of the second, which is some 12 yards below the first. Also near the end of the second groin is a depth of 5 feet, and between the second and third groins the slack water is about 4 feet in depth. The same characteristics mark the depths about and between the remaining groins, which are 6 or 7 yards apart. The deeps and slacks thus formed are keenly appreciated by the trout in times of stress, and when floods render the adjacent torrent quite impracticable it is very often possible to get a few good fish on the fly in these sheltered nooks.

And now, in this chapter, I am sorry to strike a jarring note. The quality of the trout fishing—not of the fish—has sadly deteriorated in some of our moorland rivers to which the "working-man angler" has found his way in vastly increasing numbers. I do not want to appear to take up a snobbish attitude. There is no reason why the humblest among us should not be just as good a sportsman as the highest—and he often is—but if we are to deal with plain facts there is no blinking the truth that the "working-man angler's" indifference to the best possibilities of trout fishing is, in many instances, ruining those possibilities for others. Any attempt to draw attention to the growing evil is usually met by a loud outcry in the local Press that a mere handful of selfish fly-fishers would "rob the poor man of his sport." Nothing is further from the truth, the fact being that the "poor man" is robbing not only himself, but every other frequenter of the fishery he uses, of the chances of the most delightful forms of trout fishing.

Time was, not so very long ago, when the "working-man angler"—I use the term solely because it is the recognised designation of the type of angler meant—regarded trout fishing as beyond his reach, and occupied himself for the most part with raising the catching of roach into a fine art, luring them to their fate with tempting ground bait. Then he showed a disposition to extend his range to the trout fisheries, in some instances renting trout waters, and the decent sportsman was glad that he should share the higher joys which trout fishing could offer. But did he share them? Not by a jugful! He was the old dog that was not going to learn new tricks. He simply started to do the sum in trout that formerly he had done in roach. It looked like a case of "once a sod-sitter, always a sod-sitter." Mahomet was certainly not going to trail about the river bank, or wade laboriously after the mountain, if by the mere chucking in of a generous dollop of ground bait he could bring the mountain to him. Consequently, the "working-man angler," bringing his roach-fishing methods, and tackle, to bear on the trout fisheries, rich store of blessings, in the form of fat, juicy maggots, descended in plenteous showers on the trout. They no longer had to work for their living; they were "on the dole." Who is going to spend his time in attempting to satisfy a healthy appetite by skirmishing round the attics for airy nothings when a substantial feast is laid for him in the basement? The trout in these instances have to a large extent ceased to rise to the fly, and have become bottom feeders.

In fisheries where maggot fishing is allowed it is not unusual at holiday times to land a trout so gorged with ground bait that the gentles are crawling out of its gills, and I recall more than one loud outcry in one district in the North-country that certain reservoirs, used for the storage of drinking water, were being fouled by the thousands of maggots thrown in for ground bait, while instances are not wanting of angling reports from the trout rivers to the Press being far more largely concerned with maggot fishing than any of the orthodox forms of trout angling. Here and there, intersecting carefully preserved and expensively stocked fisheries, are short lengths of practically open water, and a stroll past such spots on a bank holiday, or during the week-end, will usually reveal a congested area of bottom fishers, "roach fishing for trout," as firmly anchored to their posts as pegged-down competitors at a fishing match, shovelling in ground bait until the bed of the river must be a crawling mass of maggots. Apart from club rules or riparian owners' stipulations, and so far as the mere catching of trout is concerned, it must be left as a matter between a man and his sporting conscience whether he should ground bait for trout after the manner of him who would throw down corn to collect a covey of hand-reared game birds, and then, from behind a wall, blaze into the brown. But surely no man should, in equity, adopt means of gratifying his desires when they seriously injure his neighbour? And the worst of it is that the evil is an increasing one. Stop ground baiting for trout and the quality of the fishing on many of our moorland and Border rivers will improve appreciably—as it has deteriorated appreciably. But how it is to be stopped I am not sufficiently optimistic to suggest.

I can personally vouch for the following incident: An angler wrote for permission for a day's fishing on a certain private length, on which, it was well known, grayling were preserved as sporting fish. The owner, knowing that the applicant was an angler of note, at once accorded the permission, taking it for granted that the guest would fish in a sportsmanlike manner. Did he? Starting at the bottom of the length with what the head keeper described to me as "a sack of mixed worms and maggots," he made his way upstream, carefully baiting up each suitable swim. Then, in happy ignorance that he had been watched, he fished those swims down with the swum worm (or maggot) and departed with over forty fine grayling. He never got permission to fish that length again, and some time later he complained in the angling Press of the great difficulty experienced now in securing a day's fishing on a private length compared with the ease with which such a privilege was obtained in the good old times.

As in this case individual effort, and in other cases club restrictions, may bar the baleful practice, but intervening are stretches of easily accessible water in which the growing evil is indulged in without let or hindrance, and the degenerate influence spreads through the adjoining fisheries, to a great extent changing the natural habits of the trout and transforming them from solitary surface-feeders into gregarious gravel-grubbers.

# CHAPTER XX

## HINTS ON PRESERVATION

THE bulk of the matter which is contained in the preceding chapters may be taken as coming under the head of preservation as the term is understood to-day, but it does not include everything. There is, for instance, the question of protecting the fishery from illegal netting. With such a laudable object in view, the task of piling the river bed in the height of summer when the water is low can be confidently recommended to the personal efforts of the athletic members of a club, or the privileged friends of a private owner, as a healthy exercise little less exacting than physical jerks.

Usually the pile consists of a stout stake, wearing round its head a halo of tenpenny nails, and this should be driven into the bed of the river until the top stands some inches above the gravel, but is submerged even in low water. In considering the number and positions of the piles the disconcerting fact should not be lost sight of that, in defeating the aims of the poacher, you are also increasing your own difficulties in netting out coarse fish. Fortunately the piles are easily spotted in low clear water during the daytime, and are not so readily detected at night—at least, not until the net is hung up on them. In the former case, men knowing their whereabouts, and following behind the trammel, can, if they are not too lavishly distributed, lift the net over them. But even then accidents will occasionally occur, and these may be obviated to a great extent by the adoption of a less well-known method.

In place of the wooden stake with the nails driven into it use a long iron spike (round) of an inch or more in diameter. About an inch below the head bore a transverse hole to take, say, a $\frac{1}{4}$-inch mild steel peg, driving fit, and allow the peg to project an inch on each side. Just below this bore another hole at right angles to the first and insert another similar peg. These spikes are driven into the bed of the river until their heads, with the transverse pegs, are a few inches above the gravel. Once in position the wooden piles are awkward customers to shift, but the iron spikes, while sufficiently obstinate to discourage the attentions of poachers, can be comparatively easily lifted in daylight with the assistance of a crowbar held well to the body of the spike and below the lower peg. Thus, if it were suspected that poachers had spotted their whereabouts, they could be withdrawn and their positions changed. Such a change might well be arranged to synchronise with the raid on the coarse fish, *i.e.*, a day or two before you net take out the spikes, then net the ground while it is free from hindrance, and replace the spikes in new positions.

Should it not be deemed expedient to go to the

length and labour of removing the spikes for netting, and yet some safeguarding of the nets be desirable, then I offer the following suggestion, which theoretically seems sound, but which I admit I have never tried nor seen tried: Beg, borrow or buy common earthen plant pots of sufficient size, say about 8 inches in depth, to the number of the spikes in the length to be netted, and the day before you net place an inverted plant pot over each spike, the rim of the pot resting on the gravel. As your trammel comes along the foot of it strikes a sloping surface, and itself bellying with the current in the direction of the slope, may slip over the obstacle as it tilts forward; or it may tip it completely off the spike, in which case it tips itself harmlessly over the obstruction at the same time; in probably the very exceptional case the trammel may refuse to surmount the obstacle, in which instance no damage would be done, and all that would be required would be that the man in the river following the net should lift it over in the usual manner.

Pieces of ancient broken-meshed wire-netting, well ballasted with heavy boulders; old kettles filled with cement and liberally garnished with barbed wire, weighted iron buckets with ragged-edged, torn sides—all these have been used as wholesome deterrents to illicit netting. Once I came across the instance of a fearsome contraption of discarded scythe blades set as a *cheval de frise* in a small keg of cement and hove into the river. Care should, of course, be taken not to plant such murderous engines as this in the path of the wading angler.

Turning to another phase of preservation, the maintenance of the sporting capacity of the fishery at its highest pitch must depend largely on the conditions—club rules or private regulations—under which it is fished, and such conditions should be carefully framed with due regard to the number of rods likely to be on the water, the replenishing of the stock, naturally or artificially, the variety and proportionate numbers of the fish in the river, and so forth. A size limit for trout should certainly be imposed. In the case of the average moorland river of fair volume 8 inches will usually be found a workable standard, while on the upland tributaries the size might well be reduced to 7 inches. Whether the angler should be limited to a certain number of fish per day, and if he himself should be limited to a certain number of fishing days per season, will naturally depend on the capacity of the river in relation to the number of rods. In fisheries where grayling are fairly numerous it is often found expedient that no limit as to size or number should be imposed with regard to that fish, even in waters where it is regarded as a sporting quarry. "Vermin," of course, are vermin all the year round.

The question of lures to be permitted to the angler deserves deeper consideration than is frequently afforded it. I have on my table at the moment the draft of proposed rules for an exclusive, high-class trout fishing club which is in process of formation. The fishery concerned comprises a length of good moorland river of considerable width and volume. The first avowed object of the club is " to encourage the higher forms of angling," and to this end " the only lure allowed before June 30th shall be the artificial fly, natural fly and creeper; after June 30th worm swimming," which means to say that, in practice, artificial fly only is to be permitted, with the exceptions that advantage may be taken of the Stone-fly season, and in summer the upstream worm may be fished. No minnow fishing is to be allowed. My sympathies are entirely with the aims of the proposed club, and I merely quote from its prospectus as a concrete example of how, to my mind, the best intentioned zealot may defeat his own ends.

Say that, first and foremost, you are a fly-fisher, and your definite object is to keep the fly-fishing in the fishery in which you are interested up to concert pitch. You, therefore, interdict bait fishing. What are you going to do with the non-risers which are occupying space and absorbing precious fish food? " Fly only " may work out all right on the chalk streams where an abundance of surface food keeps the fish keenly interested in the upper shelf of the larder; but on the moorland rivers, where the hatches of fly are, for the most part, sparse and far between, the case is different. I believe it is accepted as a truism by those who have studied the habits of trout that in such waters the free risers are the fish of two, three, and, perhaps, four years old. After that age they gradually fall from grace, becoming bottom feeders and pronounced cannibals. Their life of a dozen years or so may be divided into three stages of about equal lengths. In the first they are, from the fly-fishing point of view, 100 per cent. blessings; in the second they are merely sporting half-timers; in the third they are little better than pests, refusing to look at the fly, and devouring not only the food of the risers, but the juvenile risers themselves. Fish the artificial fly only and you are draining the fishery of its risers, while the stock of older non-risers is allowed to accumulate unchecked.

For several years I made it my rule, in fishing the moorland rivers, to stick to fly until the end of June. During July, when the streams were low, I fished the clear-water worm, and during August I substituted the clear-water minnow for the worm. The variations in the results from the three methods were well defined :—

*Artificial Fly.*—Bulk of the trout about the 8-inch limit, with a considerable percentage of undersized fish returned.

*Worm.*—Bulk perhaps a trifle above the average of the fly-caught trout, with a few big fish, but still many undersized ones.

*Minnow.*—Bulk of the fish 9-, 10- and 11-inch trout, with a small percentage about the limit size. Undersized fish sufficiently rare to form a negligible quantity.

The photograph reproduced is one I happen to

WEEDING OUT THE OLD STAGERS.

[*Facing page* 152.

have of a fair day's basket with the clear-water minnow on a moorland river where the takeable fish run about four to the pound. Almost the whole of these fish were old stagers, of no use to the fly-fisher, *and not a single undersized trout was touched.*

You cannot keep a flower garden a thing of beauty and a joy for ever without constant weeding; neither can you keep the fly-fishing of a moorland river at high water mark without constantly weeding out the undesirables, no matter whether these take the form of coarse fish or trout that have outgrown the rising stage. The former you can net; the latter you can keep in check by judicious minnow fishing and by moth fishing by night in summer—the "bustard fishing" of the North-country—both of which are peculiarly fatal to the detrimental patriarchs of the river.

There is another point which should not be lost sight of by any one, private party or club trustee, proposing to rent a fishery, viz., that full possession of fishing rights can only be transferred by deed. You cannot by word of mouth fix up an arrangement with a riparian owner to take his fishing on, say, a yearly tenancy, like a house. He may by such an arrangement grant you permission to angle in his fishery, but that is all. You cannot exercise control over that fishery, and in the event of your attempting to bring a case against any offender for damage by pollution, for instance, you have no *locus standi* unless you can produce your deed leasing, or conveying, to you the rights. In blissful ignorance of this fact many angling clubs have rented water for years on slipshod arrangements until, in some case of importance, they have found themselves helpless and have been bitterly awakened to their true position. Even after the wholesome disentanglement effected by the Act of 1923 Fishery Law remains a fearful and wonderful thing.

## CHAPTER XXI

### THE RIVER-WATCHER

On the moorland river, as on the chalk stream, a good river-watcher is a tower of strength to the trout preserver—and a poor one is the very deuce. The watcher's duties go beyond touching his cap to the club members or his master's guests, and always carrying a corkscrew. He need not, however, be a Doctor of Science. True, a smattering of rudimentary entomology may at times serve him in good stead, but his job is primarily to watch the river in the interests of the owner or lessees, not scientific research. To do his job he is better equipped with a stout stick in his hand and common sense in his head—and the ability to use them when occasion occurs—than with a microscope and a familiarity with Latin names. I am writing, let it be borne in mind, of a river-watcher, not of a fishery curator.

As an illustration of the said desirable common sense let me quote a concrete example. In a certain fishery there was strong reason to suspect that night-lining was going on, but all attempts had failed to detect the culprit, or culprits, until one night the river-watcher observed a man slip out from a mill which was running night and day, and lay a night line in the river. Had the watcher yielded to impulse and accused the poacher there and then, how would the case have ended when brought into court for the purpose of making an example? No corroborative evidence; case dismissed. What actually happened was that the watcher lay doggo until the man had disappeared into the mill; then he quietly slid away and hunted up a policeman of whose whereabouts he was cognisant. The two secreted themselves until the man reappeared to lift his line, whereupon they nabbed him in the act. Result, easy conviction.

In the case of a private fishery the task of watching the river often comes within the province of the gamekeeper and his satellites; in the case of the club the separate cost frequently forms a very serious item. Still, the exercise of diplomacy and the friendly spirit may often ease that burden somewhat by enlisting the sympathies of parties whose daily callings bring them into touch with the stream. The beat of the rural policeman often touches the bank of the fishery; the members of a golf club chase the little white sphere along the other bank; game coverts border the water at another point; while elsewhere agricultural holdings extend to the riverside. Policemen, greenkeepers, gamekeepers and farmers may all be induced, by the display of a little tact, to keep an eye on the water and furnish useful reports.

Honorary watching, valuable as it may be in general, cannot be wholly depended upon, and some sort of, at least occasional, surveillance should be undertaken, otherwise—well, one friend of mine, a county magistrate, was applied to by the local policeman for leave to fish a little when he was off duty. Permission was given, my friend arguing to himself that the oftener the policeman was fishing the river the better it was being watched. His faith was rudely shattered some weeks later, when his head gamekeeper discovered that the guardian of the law, when constabulary duty was done, was not only wont to plant the fishery thickly with night lines, but had a little contract running with a hotel to supply it with trout.

Unfortunately, the professional watcher is—in rare instances, I am glad to say—too prone to lapses from strict rectitude, and it is neither undignified on the part of the employer, nor humiliating to the employed, that the master should keep a close eye on the man until he has given practical proof of his capability and fidelity. In the case of one trout fishery in which I was interested we engaged a river-watcher who seemed exceptionally well cut out for the job. He was enthusiastic and energetic. Moreover, he never objected to night duty—rather seemed to like it, in fact. We thought that we had got a treasure—until a police court case revealed to us that his energy and enthusiasm had been misdirected to ground game poaching at night on adjacent estates.

Supposing, however, that you have engaged your watcher, and that he is willing, honest and intelligent, in patrolling the fishery he will gather any information (and act upon it, or report it, as his common sense dictates) of any irregularity or suspicious happening that may have been noticed by his friends, the farmers and others. He will civilly inspect creels, tickets, etc., according to club rule or direct instructions, and he will, first, last and all the time, keep his eyes peeled for evidence of the dread presence of the preserver's arch-enemy, pollution.

Pollution is of one of two kinds. It is either incidental, spasmodic, often accidental, but none the less devastating and beyond the watcher's power of prevention. In such cases the evidence is generally not far to seek. Or it is more or less persistent, insidious, of the slow poisoning class with cumulative results that are in the end just as fatal but which are often difficult to trace to their source. Gasworks, sewage works, factories, mines—all these when in proximity to the river should be suspect as the potential origins of sudden devastation through, say, the bursting of a vat, the collapse of a bank or the wreckage of a sluice. On all these the watcher should keep an eye, and his attention should not alone be confined to his daylight patrols, for the discharge of objectionable effluents are not always the results of accidents. I have myself, while in court during the hearing of a pollution case, heard an employee of certain sewage works depose that it was a common custom at the works, when accommodation was severely taxed, to turn untreated sewage into a trout fishery after dark through what was referred to as a " back door."

The first evidence of a sudden deadly discharge of pollution into the fishery is the appearance of dead and dying fish. One dead trout may be an accident ; two dead trout in close proximity to each other are sound ground for suspicion ; three spell catastrophe in capital letters. The watcher should waste no time in critical examinations of the victims' vents and gills. He can leave that till the post-mortem. His urgent business is to track the mischief to its source and round up evidence that will be of material assistance in laying the blame at the right door. With that end in view he should press into active service the first likely witnesses he sees and proceed upstream, carefully scanning the water for fresh victims, until he can fix the spot at which the evidence ceases, and this will often be below the outfall of some effluent, above which no dead or dying fish can be found. He may, if he acts promptly, arrive at the origin of the disaster before the influx of the noxious agent ceases, and in any case he has obtained evidence which is of the highest importance.

With respect to the other class of pollution closer call is made upon the watcher's powers of observation. There is no train of dead fish to lead him to the source of the evil. It may even be that few dead trout have been observed in the river, only that the fish have gone off in condition, that hatches of fly have become more and more scanty, and that the supply of fish food has dwindled to a miserable pittance. Perhaps it has been noticed that the grayling in the length have appreciably decreased in numbers compared with the trout. In such an instance, bearing in mind that trout will continue to battle with, comparatively, mild pollution that grayling will not face, the inference is that pollution exists in sufficient strength to decimate the grayling and kill off to a great extent the aquatic creatures upon which the trout exist.

The watcher's business is to discover the source of the insidious evil, and to do this he must accustom himself to "read sign," keeping every point of influx under close observation. Perhaps there is a tributary in which formerly the spawning fish were wont to linger in leisurely fashion ; now they are in and out again with the rush of the feverish customers of the quick-lunch counter. What is the explanation ? One possible solution is that while Nature impels them up the stream to the nuptial redds, there is something in the water that inspires them with a wholesome yearning to get the nasty taste out of their mouths as quickly as possible. Or, perhaps, there is a stream into which a gutter discharges the washings from a length of sprayed macadam road. Both cases call for the most careful observation. But in the vicinity of all outfalls into the fishery let the watcher note minutely the relative conditions of things below and above the points of possible danger. Are there stains upon the stones in the river bed below the outfall, say, of a culvert that are absent above ? Is the marginal vegetation healthy above and sickly below? Is the gravel covered with slime below and not above? Is there an abundance of larvæ above and a paucity below? Do feeding fish affect the water above and eschew it below? Any results prejudicial to the tributary should be followed up by a close investigation of the stream itself, particularly with a view to establishing whether normal life exists in it or not. Then, if necessary, samples of the water may be bottled for analysis in the presence of witnesses, the samples being taken, if tar spray is suspected, during the first flush after a period of drought, otherwise when the tributary is low and the poison is, probably, least diluted.

The watcher should make it in his way to be kept posted as to sheep-dipping operations, and should make a point, if possible, of being present without making his presence obnoxious. This may often be done by lending a helping hand in the work. It gives him the dual opportunity of maintaining friendly relations with the farmer and of satisfying himself that all due precautions are taken to avoid risk of the dip finding its way into the fishery.

Where a hatchery is run in connection with the fishery the watcher will have work to do in relation to that, the nature and extent of the work depending, of course, on the staff of the fishery. It may be that the whole charge of the boxes and rearing ponds may devolve upon him, with assistance at spawning times and when the ponds are emptied. When that is so the greatest care should be exercised, before handing over the charge to him, to ascertain that he is fully competent and that his ideas on fish farming accord with those practices which experience has shown are best fitted to the peculiar features of the establishment in question. Otherwise, if he is simply to work under supervision, I am not sure that previous knowledge of the work is of great importance so long as he is intelligent and can be depended upon to carry out his instructions. Certainly he has then nothing to "unlearn," and he can be trained in the special ways of the hatchery. Also, certainly, a little learning may be a dangerous thing. I am the more strongly impressed with the truth of this since, within my recent knowledge, an enthusiastic, but misguided, assistant with but a smattering of knowledge of the work of a hatchery was, fortunately, frustrated in his efforts to dig a filter-box out of a waterway under the firm impression that he was removing a casual obstruction from a channel where a free run was desirable.

Not only the main river, but the tributary streams should receive due attention from the watcher, for it is the small becks and brooks that are the more easily poached. A small bag-net is soon fixed at a suitable spot, a length of stream above driven down, net and fish picked up, and the pirates are off and away. Nor must the netter's toll on the burn be relegated to the category of negligible trifles. I have known instances of small becks yielding upwards of 200 trout at one set of the net.

Should engineering contracts be in process of execution in the neighbourhood, such as the building of a railway or the making of a reservoir, it behoves the river-watcher to be doubly on the alert. Admirable type of British brawn and sturdy doggedness as the navvy may be, he has his little weaknesses, and one of them is fish poaching—frequently on the grand scale with a stick of dynamite. By way of variety he may call in the assistance of chloride of lime in the pursuit of his hobby. Either agent, of course, is terribly destructive within the area of its influence, killing big and little fish alike.

Long and diligent river-watching, coupled with the possession of an intelligent brain, is apt to breed in the watcher mysterious powers of intuition that are often positively uncanny. He can often, figuratively, smell the presence of danger at some point of the fishery. The psychologist may account for the phenomenon by ascribing it to some subtle process of subconscious deduction, or may even murmur something about telepathy. I offer no opinion. I only know that the possession of the faculty is a valuable asset to the river-watcher, and I have, within my own knowledge, come across one or two curious examples of its practical effect.

One case concerned the old keeper on a private estate. He occupied a cottage about half a mile from the river. One night he tumbled into bed about eleven o'clock, but could not get to sleep. He was possessed with the idea that something was wrong, and the something persistently associated itself with the river where it left the bottom corner of the park. The more he tried to assure himself that the notion was all nonsense, the more deeply it impressed itself upon him, until at length he got up and dressed. As a straight cut across the park would have made him an easy object of observation, he made a detour and came out on the river bank through a plantation, right into a nocturnal fish-netting party consisting of a neighbouring farmer and his assistants, two of the most notorious poachers of the district.

## CHAPTER XXII

### COUNTING THE COST

IN attempting to estimate the cost of preserving a trout fishery the case of the private water forming part of a sporting estate is usually so inextricably entangled, administratively and financially, with the shooting and agricultural interests that the speculative apportionment of the total expenses would be of no reliable value as a guide to those who, under dissimilar circumstances, happened to be considering the question of taking over a length of river. The keepering being under the care of those in charge of the fur and feather, and work, such as netting, that may be required on the fishery being performed by the hands on the estate during slack times, the owner-preserver finds himself in much the same position as the thrifty housewife who, in estimating the cost of the cake she had made, omitted to include the price of the butter, sugar and flour on the plea that she " happened to have those."

Consequently, to get at reliable figures we must examine the accounts of clubs and associations whose interests in the fisheries with which they are concerned are unmixed with other pursuits. Even then the figures for similar lengths of similar rivers may vary considerably according to the amount of voluntary labour expended on preservation and development by the members in such directions as piling, netting, and, if a hatchery is attached to the fishery, hunting the wild ova.

Within the last few days there has come into my hands the diary of one of the pioneer club pisciculturists who was for many years the hon. secretary of a well-known trout fishing club, and his notes on collecting spawn throw a vivid light on the lengths to which an enthusiast's devotion will carry him in a pure labour of love for the good of the fishery. Amongst his entries for 1882 are the following :—

"November 7th.—Received from A—— out of dead fish which were being cleaned for cooking 780 ova. (Part of these hatched out in 100 days.)

"November 11th.—Went to get ova, but found no hen fish ready ; nearly all cock fish that were in the becks.

"November 13th.—Went to get ova in D—— Beck and found two hens that were ready ; obtained about 370 ova.

"November 22nd.—Went to B—— Beck and obtained three ova.

"December 3rd.—Went to T—— (four or five miles away) and caught four hen fish but only obtained one ovum, being about three weeks late. From December 4th to December 15th no fresh water in the becks ; country under frost and a very heavy fall of snow, which latter began to melt and fill the river on the 16th, and on Sunday, the 17th, we noticed fish going up the beck at L——. The next day, 18th, obtained 1,200 ova from T—— G——. The beck was too big with snow broth or should probably have got more."

During that particular spawning season this enthusiast was out ova hunting, with more or less success, on no less than twenty-one days, often travelling long distances across country. In the end he collected ova from which he hatched over 21,000 fry for the benefit of the river. To-day, in the majority of cases, no such record would appear in the club annals. Instead you might find a brief item in the balance sheet: "Paid for ova from the Salmonian Fishery, £——." From beck to back kitchen the old-time amateur pisciculturist left no ovarian avenue unexplored.

The following figures are the actual amounts extracted from the accounts of well-known trout fishing clubs preserving waters of the moorland type, and representing the different classes of such organisations. The first case is that of a club renting about four miles of a river of considerable volume, whose width varies from 20 yards to 40 yards. The membership is limited to thirty-five; visitors' tickets are issued at 5*s*. per day; fly fishing and bait fishing under sporting restrictions are allowed. The year I have selected is a fairly representative one.

| Receipts: | £ | *s*. | *d*. |
|---|---|---|---|
| Members' subscriptions . . | 204 | 8 | 0 |
| Entrance fees . . . . | 42 | 0 | 0 |
| Sale of visitors' tickets, etc. . | 24 | 11 | 0 |
| Total income . . . | £270 | 19 | 0 |

| Expenses: | £ | s. | d. |
|---|---|---|---|
| Wages . . . . . | 136 | 10 | 0 |
| Rent . . . . . | 76 | 6 | 0 |
| Rates . . . . . | 40 | 2 | 6 |
| Sundries . . . . . | 4 | 5 | 1 |
| Total expenditure . . | £257 | 3 | 7 |

This club runs a hatchery, and imports Scottish ova to a liberal extent, but no references are made in the balance sheet to this part of the enterprise, for the reason that the hatchery is in a great measure self-supporting. The fishery has a few good breeding becks from which the native ova are collected by the club staff and the members, and surplus yearlings are sold to cover the cost of the imported ova.

My next example is that of a club, limited to forty members, but issuing visitors' tickets, which preserves about eleven miles of a much narrower stream than that last referred to. In the upper half the river is rapid, typical moorland water, carrying a good stock of trout and grayling, and on this portion fly fishing only is permitted for trout. On the lower water, which is more sluggish and holds a certain admixture of coarse fish, bait fishing is allowed after June 16th. The subscription is five guineas annually, with five guineas for entrance fee. Members are limited to thirty days' fishing in one season and seven and a half brace of trout in one day. During the year with which the balance sheet quoted is concerned 2,620 trout and 517 grayling were recorded as having been caught, 15,000 Scottish ova were purchased and laid down in the hatchery, and about 2,000 8-inch fish were turned into the river.

| Receipts: | £ | s. | d. |
|---|---|---|---|
| Subscriptions, entrance fees and season tickets . . . | 275 | 2 | 0 |
| Sale of visitors' tickets . . | 94 | 7 | 0 |
| Sundries . . . . . | 8 | 10 | 6 |
| Total income . . . | £377 | 19 | 6 |

The total expenditure was £407 10*s.*, the principal items of which (and the small items do not affect the present purpose) were :—

| | £ | s. | d. |
|---|---|---|---|
| Wages and employers' insurance | 213 | 13 | 3 |
| Rents . . . . . . | 57 | 11 | 6 |
| Rates and taxes . . . | 26 | 18 | 10 |
| Fish food for rearing ponds . | 10 | 16 | 10 |
| Ova . . . . . . | 11 | 12 | 6 |
| 500 8-inch trout . . . | 42 | 10 | 0 |
| Snails and caddis grubs for food stocking . . . . | 12 | 0 | 0 |
| Commission on sale of tickets . | 9 | 16 | 9 |

The next exhibit is the case of a huge amalgamation renting for the benefit of its 4,000 members one of the lower reaches of a trout river before it merges into a coarse fishery. The rights extend for six miles down one bank and three miles down the other. Any kind of legitimate fishing is allowed, and probably by far the greatest number of trout caught are taken by swimming the maggot with the assistance of ground bait. Members of the amalgamation are charged 1*s.* per day and 6*d.* per half day for fishing, while non-members may fish at double those prices if accompanied by a member. On Saturdays, Sundays and general holidays day tickets are issued at half-day prices. This length is a typical " working-man angler's " fishery. The cost of running it will be seen from the appended balance sheet :—

| | £ | *s.* | *d.* |
|---|---|---|---|
| Receipts : | | | |
| Sale of tickets . . . . | 122 | 1 | 3 |
| Grant from the Fishery Board | 33 | 0 | 0 |
| Income tax refund . . . | 1 | 16 | 0 |
| Deficiency paid out of general account . . . . . | 92 | 17 | 3 |
| Total . . . . . | £249 | 14 | 6 |
| Expenses : | | | |
| Rents . . . . . . | 120 | 0 | 0 |
| Rates . . . . . . | 42 | 5 | 1 |
| Watching . . . . . | 3 | 4 | 0 |
| Commission on sales . . . | 9 | 3 | 0 |
| Sundries . . . . . | 6 | 18 | 9 |
| Stocking with trout (1,000 two-year-olds) . . . . | 68 | 3 | 8 |
| Total expenditure . . . | £249 | 14 | 6 |

The item for watching is extremely small. Possibly it represents gratuities to gamekeepers and suchlike for keeping their eyes on the water.

Judged by the first and third cases instanced, it would seem that the cost of preserving a fairly big trout river is likely to work out at something like £60 per mile per annum, even when economic advantages, such as low hatchery expenses or cheap watching facilities, are possible. In the second instance, where the volume of water passing a certain point and the width of the river are certainly not more than half those of the other two cases, the cost per mile is roughly proportionate—£33 per mile.

The last example I quote is that of an angling club whose activities are confined to the preservation of moorland reservoirs. This, I think, is the cheapest kind of trout preservation. It is something like the cultivation of a by-product. The reservoirs are there for other, and special, purposes. It is advisable that they should be stocked with fish, and the owners are usually ready to let the fishing rights for a modest sum to responsible bodies. In the case in point the club has the fishing in several reservoirs with an aggregate area of about 300 acres. Two of these reservoirs, representing twenty-seven acres, are reserved for trout only, and the figures given refer to these trout preserves entirely. There are thirty-three members, and their limits are one day per week (exclusive of recognised holidays), three brace of 9-inch fish from one lake and not more than one additional brace from the other on the same day. The presence on the spot of a caretaker eliminates the expense of watching.

| | £ | s. | d. |
|---|---|---|---|
| Receipts : | | | |
| Members' subscriptions . . | 46 | 14 | 6 |
| Entrance fees . . . . | 6 | 6 | 0 |

| | £ | s. | d. |
|---|---|---|---|
| Donation . . . . . | 5 | 5 | 0 |
| Day tickets . . . . | 1 | 0 | 0 |
| Total income . . | £59 | 5 | 6 |
| Expenses : | | | |
| Purchase of 400 9-inch brown trout . . . . . | 40 | 0 | 0 |
| Carriage and expenses in placing fish in reservoirs . . | 6 | 9 | 3 |
| Rent . . . . . . | 5 | 0 | 0 |
| Caretaker . . . . . | 2 | 0 | 0 |
| Sundries . . . . . | 1 | 10 | 0 |
| Balance forward . . . | 4 | 6 | 3 |
| Total . . . . . | £59 | 5 | 6 |

While, owing to peculiar circumstances which may affect any fishing water, the foregoing figures cannot be held to be applicable to all fisheries, as concrete examples of what typical moorland water preservation is actually costing, they should be of value to any one entertaining the idea of taking over such a type of fishery as I have indicated, not only with regard to bargaining about rents, but when, in budgeting to cover expenses, the amount of subscription is to be fixed.

As will be seen from the foregoing figures, the rating of fishing rights now forms a serious item in angling club administration, and while these pages have been in preparation an important High Court decision has been given respecting the liability of anglers, under certain conditions, to pay rates. In Chapter XX. I drew attention to the fact that the fishing rights, in the full sense of the word, can only be transferred by deed, although in many instances—in ignorance or otherwise—the right to fish, as distinctive to fishing rights as understood in law, are often acquired on payment of an annual rent by mutual arrangement with the owner. According to the decision of a King's Bench Divisional Court, in the appeal of Mr. George S. Swayne against a decision of the justices of Longtown, Hereford, reported in *The Times* of October 30th, 1926, it would appear that, while there is no doubt that the lessee of fishing rights conveyed to him by a deed is liable to pay rates, the party who has no deed, but acquires the right to fish by the payment of a yearly sum is merely a licensee and is not liable. The Lord Chief Justice said that in the opinion of the bench the person who was only the licensee to fish was not the owner of the rights within the meaning of the Statute, for the legislature must, it was thought, be taken to have known when the Act was passed that these sporting rights were frequently made from year to year, or for a season, and therefore made special provision for this. Put in a nutshell the position is this: Take over the fishing rights by deed, you are liable to pay rates as the owner, for the time being, of the sporting rights. Acquire permission to fish without deed, you are merely a licensee and are not liable to pay rates.

# CHAPTER XXIII

## CONCLUSION

At the time in which I write we are assured that the pollution of our inland fisheries—largely through the installation of new industries which discharge noxious effluents into the rivers—is on the increase, in spite of what efforts have been put forth by fishery boards, rivers boards and other authorities. There can be no doubt that means must be found for wider and more satisfactory treatment of trade effluents, and I am sufficiently optimistic to opine that remedial measures will be devised to lessen materially, if not to remove entirely, such evils as at present exist. No good can be done by hysterical outbursts of indignation. In our saner moments it must be admitted that trade is necessary to the prosperity of the nation, while angling is a luxury, and an industry, such as beet sugar production, cannot be strangled in its infancy to suit the angry demands of a body of sportsmen.

The object of the manufacturers is not to pollute. Their aim is to carry on a business profitable to themselves and the public generally without creating a nuisance, and if, so far—as in the beet sugar industry—the difficulties are so recent of origin that reasonable time has not yet elapsed for their vanquishment, the happy consummation will be arrived at much more speedily by friendly co-operation than by the raising up of a spirit of bitter antagonism and the declaration of a state of war.

On the whole, while pollution is a thing ever to be guarded against in the trout stream, and specific cases of wholesale poisoning are not unknown, our trout fisheries have suffered from wholesale pollution far less than the lower coarse fish lengths of the rivers. This is especially so in respect to our moorland rivers and streams, which for the most part are above the danger zones of the big commercial centres. With regard to them I am convinced that we have to-day more trout water at our disposal than ever we had before, and that thanks in part to what has already been done in the way of mitigating or removing pollution.

Certain trout angling clubs which were polluted out of existence by trade effluents twenty or thirty years ago have been replaced by newer organisations which carry on successfully in the abandoned fisheries now restored to healthier conditions, while the general dereliction of the British lead mines some years ago gradually brought back, by stopping the influx of the poisonous water from the lead washing, hundreds of miles of streams to their original status as prolific trout waters. Added to these are the huge areas of artificial lakes—our Vyrnwys, our Blagdons, and scores of other extensive reservoirs—that have sprung, and are continually springing, into existence in our hilly regions. Through the extraordinary increase in the popularity of trout fishing all these available waters, new and old, are heavily taxed to supply the growing needs; and if I am wrong in my surmise—if our trout fisheries have been curtailed in extent by pollution—then, surely, that is one more reason why we should make the most of what we have left by closer preservation and the development of hitherto neglected possibilities.

There are other problems in trout preservation and the improvement of trout fisheries which cannot be dealt with in general terms. There is, for example, the unsatisfactory condition, from the angling point of view, of the English lakes. Such a subject is too big for general treatment. Much that has been said in these chapters is applicable, but beyond that the development of the potentialities of the English lakes provides a theme for close and special consideration. Moreover, however plausible the theories put forth might be for the development of the English lakes, there is no body with powers vast enough to put them into effect. I can see no chance of material improvement in that direction until the problem is made a national question.

Finally, if in this work I have appeared dictatorial or dogmatic, please blame it to the exigencies of space. After the cub reporter had been reprimanded for a too assertive style and a lack of cautious respect for the far-reaching law of libel, he turned in his next " story " something like this : " It is understood that an alleged fire broke out at, or about, No. 37, Dodger Street yesterday afternoon about, it is stated, four o'clock. The alleged conflagration, we are informed, commenced in certain premises alleged to be in the occupation of an alleged tailor, known, it is believed, as Karl Schneider," etc., etc. Lack of space would not permit me to adopt this defensive style. I have simply given verbal shape to the firm convictions that have been bred in me by a fairly long and fairly intimate acquaintance, cultivated in what leisure could be spared from a busy life, with my subject. I have no axe to grind, no professional advice to tender in exchange for a fee, no apparatus or implement to sell, and my sincerest hope is that the real service the volume renders to the man who reads it will be as great as the real pleasure it afforded to the man who wrote it.

# INDEX

# INDEX